Springer Series in Materials Science

Volume 314

The Springer Series in Materials Science covers the complete spectrum of materials research and technology, including fundamental principles, physical properties, materials theory and design. Recognizing the increasing importance of materials science in future device technologies, the book titles in this series reflect the state-of-the-art in understanding and controlling the structure and properties of all important classes of materials.

More information about this series at http://www.springer.com/series/856

Liucheng Zhou · Weifeng He

Gradient Microstructure in Laser Shock Peened Materials

Fundamentals and Applications

ZHEJIANG UNIVERSITY PRESS
浙江大学出版社

Liucheng Zhou
Science and Technology on Plasma
Dynamics Laboratory
Air Force Engineering University
Xi'an, Shaanxi, China

Weifeng He
Science and Technology on Plasma
Dynamics Laboratory
Air Force Engineering University
Xi'an, Shaanxi, China

ISSN 0933-033X   ISSN 2196-2812   (electronic)
Springer Series in Materials Science
ISBN 978-981-16-1749-2   ISBN 978-981-16-1747-8   (eBook)
https://doi.org/10.1007/978-981-16-1747-8

Jointly published with Zhejiang University Press, Hangzhou, China
The print edition is not for sale in China Mainland. Customers from China Mainland please order the
print book from: Zhejiang University Press.

Foreword

The surface integrity of metal components significantly influences their service performances and lives. Fatigue fracture, for instance, has been the common result of the crack initiation from the imperfect surface and further propagation throughout the component. Surface strengthening by advanced processing technologies is hence highly expected. Laser shock peening (LSP) was developed more than half-century ago and has gained wide applications in the aviation industry. It utilized nanosecond laser pulse to induce ultra-high-pressure shock wave on the surface of the materials under water confinement, which leads to gradient microstructure evolution and gradient compressive residual stress in the processing area and significant improvement of the fatigue performance of the processed materials.

This book is completed by joint efforts of Dr. Liucheng Zhou, Dr. Weifeng He and other scholars. They have studied the theories and techniques of LSP for more than ten years, and their findings have already been applied in strengthening and repairing of multi-type aero-engine components successfully, which provides strong theoretical and practical basis of the book. On the one hand, this book emphasizes the innovation in both theory and method, and systematically proposed the formation mechanism and regulating method of gradient microstructure induced by LSP, revealing the mechanism of delaying crack initiation and propagation, the constitutive modelling, strengthening–toughing mechanism, vibration suppression mechanism, thermal and loading stability mechanism of gradient microstructures, etc. These findings enrich and extend the theory of LSP, broadening its research and application scope at the same time; on the other hand, this book focuses on solving practical engineering problems, raising many new LSP techniques such as small scanning-spot LSP, wave-transparent LSP, unequal energy density LSP and high-temperature-assisted LSP. The research has accumulated a large amount of experimental data, and the findings have been applied in aero-engines in service. It can be anticipated that the book will play an important role in solving the problems of damaged fan compressor blade by foreign object, fatigue fracture of high-temperature components and premature failure of cladding and repairing parts, etc.

Nanjing, China

Shan-Tung Tu
Professor, East China University
of Science and Technology;
Member, Chinese Academy
of Engineering

Preface

The structural components of aviation equipment usually operate under an extremely harsh and complicate environment, and its safety in service was critical. The fatigue failure of parts affects the security of aircraft directly, which is the leading killer among the flight accidents. Generally, fatigue fracture proceeds by the initiation of fatigue cracks on the component surface under the effects of cyclic loading and further propagation throughout the component up to its fracture.

To improve the reliability of components in operation and extend the service life, surface strengthening techniques have been extensively investigated and applied under the condition of having no changing mechanical properties of basal materials. Laser shock peening (LSP) technique is a typical surface strengthening process that has been widely used in aviation equipment due to its advantages such as better effectiveness, deeper affected layer, strong controllability and great adaptability. It utilizes high-energy laser with the pulse duration of several ns to irradiate the surface materials, inducing high-pressure shock wave (GPa level) propagating into the materials. Under the mechanical effects of the shock wave, the surface materials experience severe plastic deformation, leading to the formation of gradient microstructures and gradient compressive residual stress. In this way, the mechanical performances such as anti-fatigue, anti-wear and anti-corrosion of the processed materials are significantly improved.

During the investigations about the fatigue failure of components in aero-engine, it was noticed that the fatigue problems of blade and weldment are very serious, accounting for ~40% among the malfunctions of aero-engine. For example, the fan/compressor blades are frequently damaged by foreign objects, which has been a prevalent problem in aero-engine. During taking off and landing of aircraft, some hard objects such as sand, cement and metals would be inevitably inhaled into the aero-engine, causing the damage of fan/compressor blade and reducing their fatigue performance. In addition, in advanced aero-engine, welding structures are widely used in the thin-walled structural parts and pipes. However, due to the thermal effects during welding process, massive brittle phases and coarse grains as well as tensile residual stress are formed in the heat-affected zone of weldments, which results in vibration fatigue failure more easily. To solve these problems, we systematically investigate the effects of LSP on these structural components which are prone to fatigue failure. We unite Xi'an Jiaotong University, Southwest Jiaotong University, Xi'an Tyrida Optical Electric Technology Co., Ltd and so on, focusing on fundamental research and systematically proposing the anti-fatigue theories and techniques of LSP-induced gradient microstructure. Now, LSP techniques have been successfully applied in ten kinds of aero-engine parts such as blades, pipes, gears, casing welds, flame tubes and small holes. Besides, it has also been used in the strengthening of ground gas turbine blades, nuclear reactor welds, aircraft pipes, bicycle frames, etc.

This book systematically summarizes the research results of our research team on the LSP-induced gradient structure and refers to relevant literatures at home and abroad. The book is divided into seven chapters: Chap. 1 introduces the basic principles and development status of LSP and the reason why we study the LSP-induced gradient structure; Chap. 2 introduces the pressure model and propagation characteristics of laser-induced shock waves; Chap. 3 introduces the formation mechanism and techniques of LSP-induced gradient structure of titanium alloy and steel; Chap. 4 introduces the anti-fatigue mechanism of the LSP-induced gradient structure of titanium alloy and introduces the related research for improving the damage tolerance of the blade damaged by foreign object; Chap. 5 introduces the anti-fatigue mechanism of the LSP-induced gradient structure of the nickel-based superalloy, the thermal stability mechanism of the gradient structure under high-temperature environment and the processing method; Chap. 6 introduces the microscopic mechanism of the LSP-induced gradient structure to suppress crack propagation and the crystal plastic constitutive model of the gradient structure, and explores the mechanism of the LSP-induced gradient structure to suppress the vibration characteristics of thin-walled components; Chap. 7 mainly introduces the compound mechanism and processing methods of LSP and other surface technologies such as vibratory finishing, shot peening and surface infiltration. It also introduces the compound process of LSP and additive manufacturing, welding, etc., which could improve the fatigue strength of repairing components.

This book is written by myself, Weifeng He, Xinlei Pan, Xuede Wang, Xiangfan Nie, Sihai Luo, Lingfeng Wang and others. In the research work involved in this book, Prof. Yinghong Li participated in the guidance throughout. In addition to the above-mentioned teachers, Prof. Li Cheng, Dr. Jingyuan Liu from Air Force Engineering University, Prof. Xu Zhang, Prof. Cai Zhenbing, Prof. Qianhua Kan from Southwest Jiaotong University and others also participated in related research. I would like to express my sincere thanks to these colleagues, comrades and classmates!

This book is supported by the National Science and Technology Major Project of China (grant numbers J2019-IV-0014-0082, 2017-VII-0003-0096 and 2016YFB1102600), National Natural Science Foundation of China (grant number 51875574), the Youth Talent Promotion Project of China (grant number 17-JCJQ-XX) and so on. Thanks for the above-mentioned funding of the project!

Special thanks to Academician Shan-Tung Tu for his preface to this book!

Due to our limited knowledge and energy, there inevitably exist some ambiguous interpretations and even mistakes in this book, which we welcome the readers and colleagues to point out.

Xi'an, China   Liucheng Zhou

Contents

Contributors

Liucheng Zhou Science and Technology on Plasma Dynamics Laboratory, Air Force Engineering University, Xi'an, Shaanxi, China

Weifeng He Science and Technology on Plasma Dynamics Laboratory, Air Force Engineering University, Xi'an, Shaanxi, China

Xinlei Pan Science and Technology on Plasma Dynamics Laboratory, Air Force Engineering University, Xi'an, Shaanxi, China

Xuede Wang Science and Technology on Plasma Dynamics Laboratory, Air Force Engineering University, Xi'an, Shaanxi, China

Xiangfan Nie Science and Technology on Plasma Dynamics Laboratory, Air Force Engineering University, Xi'an, Shaanxi, China

Sihai Luo Science and Technology on Plasma Dynamics Laboratory, Air Force Engineering University, Xi'an, Shaanxi, China

Lingfeng Wang Science and Technology on Plasma Dynamics Laboratory, Air Force Engineering University, Xi'an, Shaanxi, China

Chapter 1
General Introduction

Under the action of alternating load, metal parts are often scrapped thoroughly due to local fatigue cracks and fractures, and even lead to devices or equipment failure or accidents [1]. According to statistics, in the aviation field, 40% of the mechanical failures of aircraft and metal engine parts are caused by fatigue. Because the maximum working stress and the initiation of fatigue cracking are generally located on the surface of a component, in order to improve the reliability of structures under service conditions and prolong their service life, surface strengthening technology has been studied increasingly without changing the properties of substrate materials, and various surface strengthening methods have been put forward, which have brought about good results and reaped some benefits [2]. Laser shock peening (LSP) technology has been widely discussed and studied because of its advantages of a good strengthening effect, strong controllability and good adaptability, and it has been successfully used in the aviation field to improve the fatigue performance of metal parts [3–5]. Most of the existing research focuses on the simulation, optimization of the residual stress field formed by laser shock wave and the mechanism for the improvement of fatigue performance. Under certain conditions, laser shock wave can form gradient nanostructures, which have better stability of loading under fatigue conditions and in a hot environment, and can significantly improve the fatigue performance of metal parts.

Aiming at titanium alloy and Ni-based alloy parts which are widely used in the aviation field, this book introduces the characteristics of the gradient structure and the mechanism of the formation of two kinds of metal materials under the action of a laser shock wave, the stability of the gradient structure under the action of heat, the mechanism of the strengthening and toughening of the gradient structure and its influence on fatigue performance, and the composite application of LSP and other surface technologies.

© Zhejiang University Press 2021
L. Zhou and W. He, *Gradient Microstructure in Laser Shock Peened Materials*, Springer Series in Materials Science 314,
https://doi.org/10.1007/978-981-16-1747-8_1

1.1 Typical Applications of LSP in the Aviation Field and Recent Development

The basic principle of LSP is that high-power (GW/cm^2 magnitude) and short-pulse (ns magnitude) laser irradiates the surface of metal material, and the absorbing layer (aluminum foil or black adhesive tape) coated on the metal surface quickly absorbs the laser energy and explosively gasifies, and instantly generates plasma. Under the influence of the water confinement layer, the recoil of plasma gas flow forms a high-pressure shock wave of a GPa magnitude, which propagates into the material. When the shock wave pressure is greater than the strength of the dynamic yield of the material, the material undergoes plastic deformation at an ultra-high strain rate (>10^6/s), as shown in Fig. 1.1. Compared with other laser surface treatment technologies, LSP technology is characterized by utilizing the "force" effect of the shock wave, that is, LSP uses the "force" effect of a laser-induced plasma shock wave to form a gradient structure of residual compressive stress and microstructure on the surface of metal materials, thus improving the fatigue resistance [3, 6, 7], wear resistance [8–10], stress corrosion resistance [4], etc.

LSP has three main characteristics:

(1) High pressure. The pressure of the detonation wave reaches GPas, even TPas. This is difficult to achieve by conventional machining. For example, the pressure of mechanical stamping is often between tens of MPa and hundreds of MPa.

(2) High energy. The single pulse energy of the laser beam reaches several tens of joules (J), and the peak power can reach GW level, which can convert light energy into shock wave mechanical energy within 10–30 ns, thus realizing the efficient utilization of energy.

(3) Ultra-high strain rate. The action time of the shock wave is only tens to hundreds of nanoseconds, and the strain rate reaches 10^6 s^{-1}, which is 10,000 times

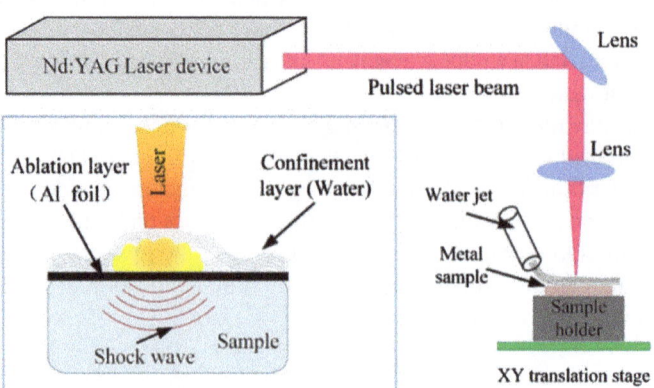

Fig. 1.1 Schematic diagram of LSP

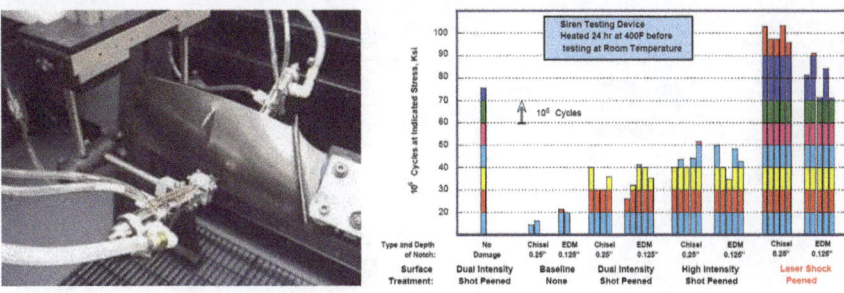

Fig. 1.2 LSP and damage tolerance data of the primary fan blade of an F101 engine

higher than that of mechanical stamping and 100 times higher than that of explosive forming. Therefore, it is a manufacturing method under extreme conditions in an extreme environment.

Because LSP can effectively improve the fatigue strength of materials, many units have carried out research and jointly promoted its technical progress from different aspects. Up to now, many international academic conferences on LSP have been held. At present, the application of LSP mainly focuses on aero-engine fan/compressor blades, turbine blades, integral blades, disk shafts, ducts, aircraft structural parts (holes, edges and lugs, etc.), helicopter transmission gears, carrier-based aircraft hooks, nuclear waste container welds, nuclear fuel rod shell welds, blades of large-scale gas turbine/water turbine, and key components of automobiles, etc.

In 1995, the establishment of the Laser Shock Peening Technology Company (LSPT) marked the fact that the LSP technology had begun to enter the engineering application stage. In 1997, GE Company used LSP technology to increase the damage tolerance of the leading edge of the Ti811 titanium alloy blade of the primary fan of F101 engine by 1500%, which effectively solved the problem of the reduction of fatigue strength caused by foreign objects in the cold end blade of the aero-engine, as shown in Fig. 1.2 [11].

In 2004, in the "Engine Structural Integrity Outline" (ENSP-1783B-Change 2) issued by the US Department of Defense, it was explicitly proposed to incorporate LSP technology into the structural design and manufacture [12] of engines, and at the same time, the LSP process specification ANSI Z136.1 and standard AMS 2546 were issued. It shows that the technology has reached the stage of engineering application and can effectively support the design, development, manufacture and repair of aircraft and engines. At present, the United States has successfully applied LSP to F101, F110, F404 and other engine fan/compressor blades equipped on the third/fourth generation aircraft, effectively solving the problem of insufficient anti-foreign object injury ability of aero-engine fan blades, and then introduced it to F119 and F135 engines installed on F-22 s and F-35 s. The number of strengthened blades reaches 30,000–40,000 pieces every year. Figure 1.3 shows the application of LSP technology in different types of aircraft and the engines they are equipped with in the United States by 2015.

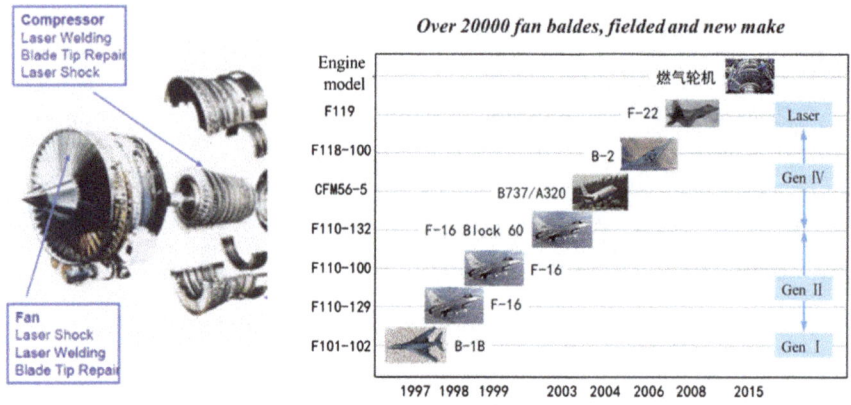

Fig. 1.3 Application of LSP technology in aircraft/engine components in the USA

In 2003, the Federal Aviation Administration (FAA) and Japan Asia Airways (JAA) approved LSP technology as the maintenance technology of the key components of aircraft/engines, and introduced it to the blade manufacturing and maintenance of the CFM-56 engine installed on Boeing 737 s and the GE90 engine installed on the Airbus A380. The Metal Improvement Company (MIC) established an LSP production line at the Rolls-Royce Company, in England, to strengthen the blades of the Trent series of engines [13]. Since 2005, the United States has continued to expand LSP technology to the treatment of key components of gas turbines, petrochemicals and automobiles. With the continuous development of technology, it was further introduced to the hook of carrier-based aircraft and the bolt hole of aircraft for wing-body connections [14], which effectively solves the problem of failure due to fatigue. After 2008, MIC established an LSP production line in Earby, England, which is mainly used to strengthen the fan blades of the Rolls-Royce Trent 500, Trent 800 and Trent 1000 engines to improve their ability to prevent high-frequency fatigue and foreign object damage. In 2017, LSPT Company launched the Procudo® LSP system, which has a good effect in improving working efficiency and equipment stability, as shown in Fig. 1.4.

Fig. 1.4 Procudo® laser peening system

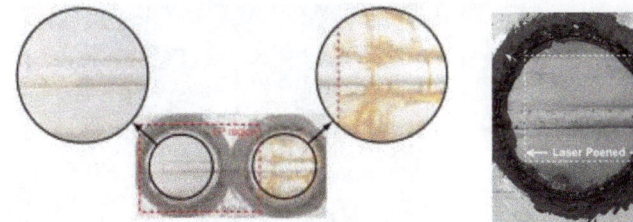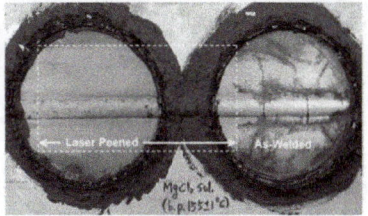

Fig. 1.5 Enhancement of stress corrosion resistance of metals by LSP (the picture on the right is enlarged)

In the field of welding seam treatment, LSP is used for welding manufacturing and damage repair of engine parts, eliminating residual tensile stress, preventing stress corrosion and improving fatigue strength [15–17]. In addition, from 2000 to 2002, the Lawrence Livermore National Laboratory in the United States verified that LSP can greatly reduce the stress corrosion and crack growth rate of welds; the nuclear waste storage containers in the United States are buried deep under Yucca Mountain, and they must be stored for 10,000 years without leakage. However, due to the residual tensile stress in the welds of the containers, cracks initiate, expand and accelerate corrosion; experimental studies have proved that LSP can greatly reduce stress corrosion and the growth of cracks in the welds, and this technology has been applied to the strengthening of welds on nuclear waste containers (Fig. 1.5).

Toshiba Corporation of Japan has studied the laser shock processing technology and has developed a set of unique LSP processing equipment and technologies. The welding seams of nuclear reactor pressure vessels and pipeline joints were treated by LSP with small energy and in small spots (spot φ 0.8 mm, energy 200 mJ, pulse width 8 ns), in order to improve the resistance to stress corrosion cracking of the welding seams.

In addition, a new beam deck delivery system (BDDS) proposed by the LSPT Company will use LSP to treat the aluminum deck section of warships to enhance the fatigue strength of metal. At present, LSP technology is widely used in the maritime field, mainly including the repair and improvement of hulls, propellers, engines and propulsion systems, shafts, bilge keels and other parts. LSP can also be customized to correct deformation and form compound curvature, as shown in Figs. 1.6 and 1.7.

As for the civil aviation engine, LSP technology was approved by FAA and JAA as the maintenance technology of aircraft/engine key parts in 2003, and it has been applied to the manufacturing and repairing of engine blades such as the CFM-56 on Boeing 737 aircraft and the GE90 on the A380 aircraft. After 2005, it was gradually introduced to large gas turbines (e.g., 7F gas turbines of GE), to the blade treatment of water turbines, and to the treatment of key parts of automobiles, which achieved great economic benefits.

China has also carried out more research on the application of LSP technique. In 2007, Air Force Engineering University, in cooperation with Xi'an Tyrida completed the R&D, manufacturing and the development of an application software system of the first complete set of LSP equipment in China, and built the first industrial

Fig. 1.6 Analysis of the deformation correction and finite element modeling of LSP

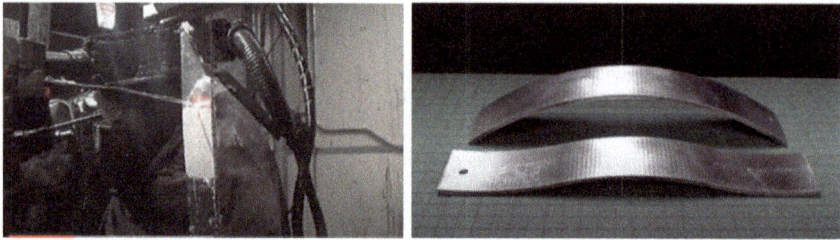

Fig. 1.7 LSP treatment of an aircraft wing panel by LSPT Company

demonstration production line of LSP in 2008 [18–21], which laid a foundation for the engineering application of LSP technology in China. Xi'an Tyrida Optical Electronic Technology Co., Ltd. is the only high-tech enterprise specializing in the research, production, sales and processing services of complete sets of LSP devices in China. The company has successively undertaken LSP machining services and experimental tasks for accessories and blades of more than 10 models of 7 series of aero-engines and ground gas turbines. After passing GE Energy Company's examination on the first workpiece treated by LSP (where the key indexes such as residual compressive stress and fatigue life have reached the leading international level), it has become the only overseas enterprise approved to provide LSP services for ground gas turbine blades of GE Energy (Fig. 1.8).

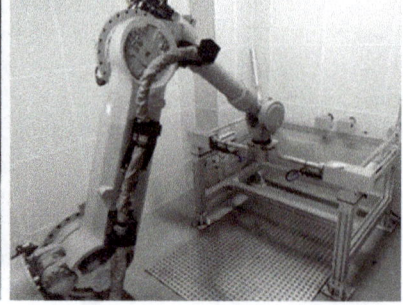

Fig. 1.8 LSP system

In addition, Jiangsu University, Beijing Aeronautical Manufacturing Technology Research Institute and Shenyang Institute of Automation of the Chinese Academy of Sciences have also carried out engineering applications of the LSP technology.

1.2 Why Do We Study the Gradient Microstructure Induced by LSP in Aeronautical Materials?

LSP can form residual compressive stress and a gradient structure on the surface of metal materials, and improve fatigue performance. Residual compressive stress can improve the fatigue strength of materials mainly by reducing the average stress borne by components, reducing the rate of the growth of cracks and even the closing of cracks. P Peyre and R Fabbro described the mechanism for the formation of residual stress of LSP [22, 23], as shown in Fig. 1.9: (1) when a shock wave is loaded on the surface of a metal material, uniaxial stress and plastic deformation will occur along the propagation direction of the shock wave; (2) after the shock wave, the plastic deformation area is limited by the surrounding metal materials, and a compressive stress field is generated parallel to the impact surface. This view is generally accepted.

It is known from a large part of the literatures that the research on improving fatigue strength of metal materials by LSP is generally based on the strengthening mechanism of residual compressive stress, focusing on the magnitude and depth of residual compressive stress. However, metal parts in the aviation field generally work under high temperatures and high pressure, and the residual compressive stress formed by LSP will mostly relax in a high temperature environment and a mechanical environment, which will affect the strengthening effect. The development of precision manufacturing on the surface of micro-nano structures and the realization of a high-strength and high-toughness design of materials are effective means for improving

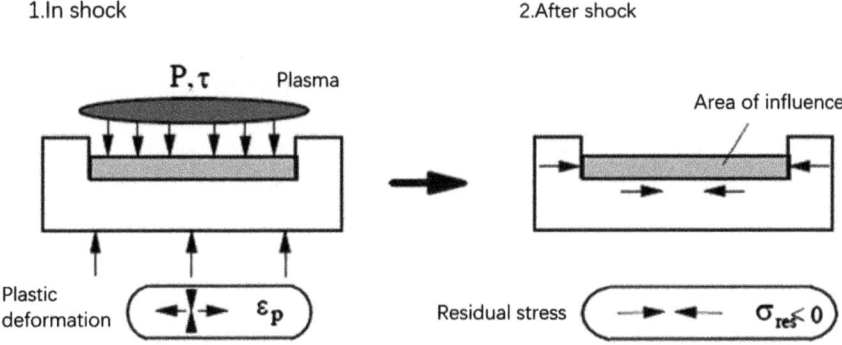

Fig. 1.9 Schematic diagram of residual compressive stress induced by LSP

the working performance of parts in an extreme service environment. The strain rate induced by a laser shock wave in metal materials is extremely high, which will form a gradient structure in the materials. Compared with residual compressive stress, LSP-induced gradient microstructure has better stability in an extreme service environment [5], which plays an important role in improving fatigue performance.

The preparation of gradient structures on metal surfaces by LSP has become a hot research topic. These studies include the behavior of plastic deformation and grain refinement of materials under an ultra-high strain rate [5, 24–26], model simulation and experimental observation of dislocation induced by the laser shock wave through molecular dynamics [27], nano-precipitation behavior under dynamic hardening induced by a shock wave [28], and the mechanism of improving fatigue performance by gradient structure, etc. Clauer et al. [11] used the shock wave induced by nanosecond pulsed laser to induce high dense dislocations on the surface of the 7075 aluminum alloy to improve its mechanical properties. Fairand et al. [29] carried outLSP tests on the 6061-T6 and the 2024 aluminum alloys, and found that these materials produced a substructure after LSP treatment, and their hardness and strength were greatly improved. Altenberger et al. [30] of Lawrence laboratory in the United States studied the effects of rolling and LSP on the changes in microstructure and the thermal stability of the 304 stainless steel and the TC4 titanium alloy. It was found that rolling can produce surface nanocrystals on the surface of the stainless steel and titanium alloy, which has good thermal stability and increases the temperature of the critical growth of both materials. In addition, LSP can produce high dense dislocations in the material, which has better thermal stability compared with that produced by rolling treatment, and LSP raises the recrystallization temperature of the 304 stainless steel and of the titanium alloy to 800 °C and 900 °C respectively. Yanushkevich et al. [31] used LSP to strengthen stainless steel, and found that the shock wave had a selective scattering propagation at grain boundary, which clears away crystal lattice impurities and makes the structure uniform, thus improving corrosion resistance and causes martensitic transformation. Cao et al. [32] o studied the influence of LSP on the substructure defects of single crystal copper. They found that when the shock wave pressure is in the range of 55–60 GPa, micro twins are formed along the {100} lattice direction. Fan et al. [33] established the simulation model of laser shock forming (LPF), and measured the deformation results by the contour method. They revealed the changing process of shock wave-solid interaction. Peralta et al. [34] of the Department of Mechanical and Aeronautical Engineering of Arizona State University used a laser to strengthen single crystal NiAl, and the pressure of the shock wave was about 15 GPa. It was found that a large number of dislocation structures and lattice rotations were formed in the peening zone, and the lattice rotated and attenuated rapidly in the direction of {100} in the surface layer. Cheng et al. [12] of the Department of Mechanical and Material Engineering of Washington State University used the laser shock wave to strengthen the brittle materials of monocrystalline silicon, and proposed the theory of multi-scale dislocation dynamics to explain the strengthening mechanism. Mordyuk et al. [35] used a multi-time LSP with a power density of 0.35 GW/cm^2 to strengthen AISI321 steel, and found that dislocation walls and dislocation cells with a large angle were formed in the material, and the depth of affected layer was about 10 μm, where the high dense dislocation improved the fatigue performance of AISI321 steel. Ye et al. [36] carried out LSP experiments on heated parts, and put forward the concept of Warm Laser Shock Peening (WLSP). Under the effects of heat, AISI4140 steel will undergo dynamic strain aging in response to shock waves, and nano-precipitates will be produced through the pinning effect of carbon atoms on dislocations, which will increase the density and stability of dislocation of steel after WLSP treatment, thus improving the fatigue performance under the conditions of a hot environment. Then they studied the LSP of AISI304 stainless steel at − 196 °C. The decrease in temperature is more beneficial to the phase transformation and twinning deformation of 304 stainless steel. Lu et al. [37, 38] carried out LSP tests on metal materials such as stainless steel and aluminum alloy, and analyzed the grain refinement process. High-density deformation twins were formed in the 304 stainless steel after LSP. Increasing impact times can change the direction of deformation twins, make them cut each other and refine the crystal grains. The process of the forming of crystal grains of aluminum alloy is as following: LSP induces high-density dislocation, forms a dislocation wall, then forms sub-grain and then smaller crystallite. Chen et al. [39] treated laser welded Incoloy 800H with LSP, and then analyzed the effect of LSP on microstructure and residual stress of weldments by optical microscope, TEM and X-ray diffraction (XRD), and evaluated the effect of LSP on the mechanical properties of weldments by microhardness. The results show that the columnar grains are produced in the welding zone due to the high heat input during the welding process, and there were very few equiaxed grains. The grain refinement occurred in the peening zone, and the original lath structure was refined into equiaxed grains. The plastic deformation at a high strain rate resulted in dislocation movement (slippage), which accumulated on the strengthened surface and subsurface, and the dislocation density increased significantly.

Chandrasekar et al. [40] studied the effects of LSP on dissimilar weledments between the Inconel600 plate and the AISI316L plate joint by Activating Tungsten Inert Gas (ATIG) method, and tested the mechanical properties of the strengthened welded joints, as shown in Fig. 1.10. The results show that tensile failure occurs in the welding zone due to coarse grains and intermetallic compounds. After LSP, the residual tensile stress on the weldment is transformed into residual compressive stress, the tensile strength is increased from 573 to 630 MPa, and the resistance to corrosion is improved.

Reddy et al. [41] studied the influence of LSP (without a protection layer) on improving the metallurgical and mechanical properties of C-276 welded joints, and evaluated the transformation of the phase and the grain size in the fusion zone by means of X-ray diffraction analysis, as shown in Fig. 1.11. The results show that it is easy for the micro-segregation of alloy elements to produce hot cracks in weldments, and the formation of the topological close pack phase is the main cause of hot cracks. Fine equiaxed dendrites are observed under both conditions. EDS analysis shows that there is no secondary intermetallic phase, and the grain size decreases by 49% after LSP. Compared with welded samples, the hardness and tensile strength of laser shock peened (LSPed) samples are increased correspondingly.

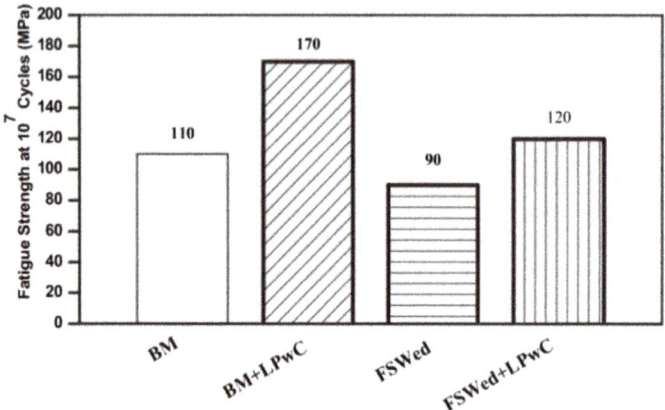

Fig 1.10 Fatigue strength of welding samples under different treatment conditions

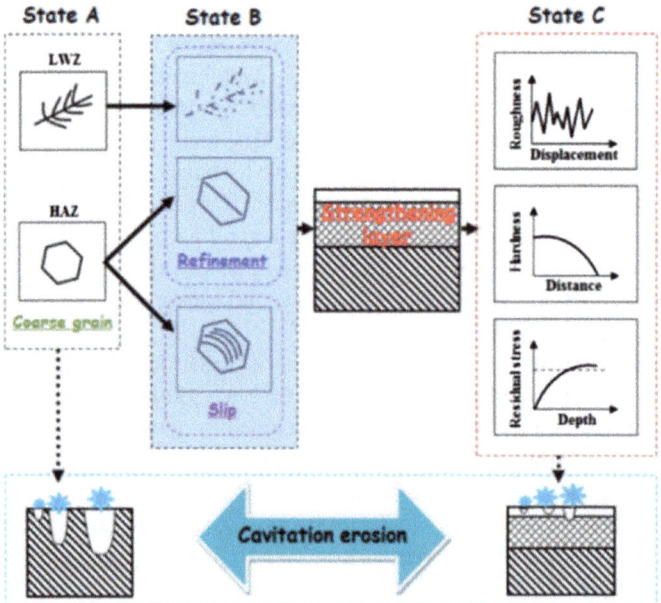

Fig. 1.11 Schematic diagram of LSP improving the cavitation resistance of welded joints

Li et al. [20, 21] put forward the concept of LSPed surface nanocrystallization aiming at the strengthening of high-temperature turbine blades, holding that when the pressure of the laser-induced shock wave is greater than a certain threshold, it can produce nanocrystals on the surface of metal materials. They successfully prepared gradient nanostructures on Ni-based alloy turbine blades of some types of aero-engines, which solves the problems of relaxation of residual compressive stress at

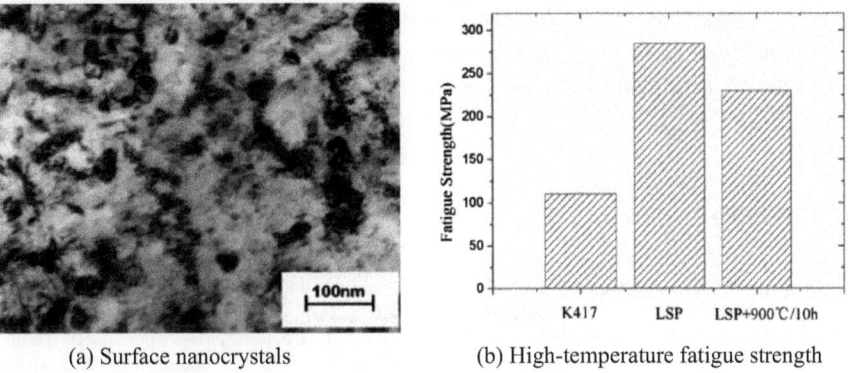

(a) Surface nanocrystals   (b) High-temperature fatigue strength

Fig. 1.12 The surface nanocrystals of K417 Superalloy were formed by LSP and the high temperature fatigue strength was improved

high temperatures and with a limited strengthening effect, effectively improves the fatigue strength of turbine blades at 800 °C and breaks through the temperature limit of 538 °C of the Ni-based alloy in the AMS2546 specification, as shown in Fig. 1.12.

1.3 Scope of the Book

LSP technology involves many disciplines, such as laser, materials, mechanics, plasma science and so on. This book takes titanium alloy and Ni-based alloy, which are widely used in aviation materials, as the research objects, and introduces the effect of the improvement of the fatigue performance of gradient structures induced by LSP. The main contents include:

(1) The model of the modification of laser shock wave pressure. Laser shock wave is the energy carrier that produces residual stress and gradient structure. This book introduces the basic principle and characteristics of laser plasma, and puts forward a new laser shock wave pressure estimation model based on experimental testing;

(2) The mechanism of the formation and the parameter range of gradient structures induced by LSP, the stability of gradient structure under heat and load, the mechanism and the effect of improving fatigue performance of the gradient structure of TC4, TC6 and TC17 titanium alloys;

(3) The mechanism of the formation and the parameter range of gradient structures induced by LSP, the stability of gradient structures under heat conditions and the effect of improving the high-temperature fatigue performance of GH4133B and K417 Ni-based alloys;

(4) The mechanism of the strength-toughening of gradient structures and the improving effect on vibration characteristics of thin-walled structures;

(5)   The effects on fatigue performance of aero-engine blades by LSP combined
      with other surface treatment technologies, such as LSP and shot peening, LSP
      and additive manufacturing, LSP and the aluminizing process, etc.

References

1. S.A. Namjoshi, S. Mall, Fretting behavior of Ti-6Al-4V under combined high cycle and low
    cycle fatigue loading. Int. J. Fatigue **23**, S455–S461 (2001)
2. G.H. Majzoobi, K. Azadikhah, J. Nemati, The effects of deep rolling and shot peening on fretting
    fatigue resistance of Aluminum-7075-T6. Mater. Sci. Eng. A-Struct. Mater. Prop. Microstruct.
    Process. **516**(1–2), 235–247 (2009)
3. X. Pan, S. Guo, Z. Tian, P. Liu, L. Dou, X. Wang, Z. An, L. Zhou, Fatigue performance
    improvement of laser shock peened hole on powder metallurgy Ni-based superalloy labyrinth
    disc. Surf. Coat. Technol. 126829 (2021)
4. C.Y. Wang, K.Y. Luo, X.Y. Bu, Y.Y. Su, J. Cai, Q.L. Zhang, J.Z. Lu, Laser shock peening-
    induced surface gradient stress distribution and extension mechanism in corrosion fatigue life
    of AISI 420 stainless steel. Corros. Sci. **177** (2020)
5. X.K. Meng, H. Wang, W.S. Tan, J. Cai, J.Z. Zhou, L. Liu, Gradient microstructure and vibration
    fatigue properties of 2024-T351 aluminium alloy treated by laser shock peening. Surf. Coat.
    Technol. **391** (2020)
6. M. Kahlin, H. Ansell, D. Basu, A. Kerwin, L. Newton, B. Smith, J.J. Moverare, Improved
    fatigue strength of additively manufactured Ti6Al4V by surface post processing, Int. J. Fatigue
    134 (2020)
7. X.L. Pan, X. Li, L.C. Zhou, X.T. Feng, S.H. Luo, W.F. He, Effect of residual stress on S-N
    curves and fracture morphology of Ti6Al4V titanium alloy after laser shock peening without
    protective coating. Materials **12**(22), 12 (2019)
8. Y. Guo, S.R. Wang, W.T. Liu, Z.L. Sun, G.D. Zhu, T. Xiao, Effect of laser shock peening on
    tribological properties of magnesium alloy ZK60. Tribol. Int. **144** (2020)
9. J.Z. Zhou, Y.J. Sun, S. Huang, J. Sheng, J. Li, E. Agyenim-Boateng, Effect of laser peening
    on friction and wear behavior of medical Ti6A14V alloy. Opt. Laser Technol. **109**, 263–269
    (2019)
10. A. Siddaiah, B. Mao, Y.L. Liao, P.L. Menezes, Surface characterization and tribological
    performance of laser shock peened steel surfaces. Surf. Coat. Technol. **351**, 188–197 (2018)
11. A.H. Clauer, B.P. Fairand, Interaction of laser-induced stress waves with metals. Appl. Lasers
    Mater. Process. (1979)
12. G.J. Cheng, M.A. Shehadeh, Multiscale dislocation dynamics analyses of laser shock peening
    in silicon single crystals. Int. J. Plast. **22**(12), 2171–2194 (2006)
13. M. Jean-Eric, B. Gérard, Laser generation of stress waves in metal. Surf. Coat. Technol. **70**(2–3),
    231–234 (1995)
14. B.S. Holmes, W.E. Maher, R.B. Hall, Laser-target interaction near the plasma-formation
    threshold. J. Appl. Phys. **51**(11), 5699–5707 (1980)
15. A.F.M. Arif, Numerical prediction of plastic deformation and residual stresses induced by laser
    shock processing. J. Mater. Process. Technol. **136**(1–3), 120–138 (2003)
16. W.J. Jia, H.Z. Zhao, Y.X. Zan, P. Guo, X.N. Mao, Effect of heat treatment and laser shock
    peening on the microstructures and properties of electron beam welded Ti-6.5A1–1Mo-1V-2Zr
    joints. Vacuum **155**, 496–503 (2018)
17. B. Dhakal, S. Swaroop, Review: laser shock peening as post welding treatment technique. J.
    Manuf. Process. **32**, 721–733 (2018)
18. X. Nie, W. He, L. Zhou, Q. Li, X. Wang, Experiment investigation of laser shock peening
    on TC6 titanium alloy to improve high cycle fatigue performance. Mater. Sci. Eng. A **594**,
    161–167 (2014)

19. X. Nie, W. He, S. Zang, X. Wang, J. Zhao, Effect study and application to improve high cycle fatigue resistance of TC11 titanium alloy by laser shock peening with multiple impacts. Surf. Coat. Technol. **253**, 68–75 (2014)

20. S. Luo, L. Zhou, X. Nie, Y. Li, W. He, The compound process of laser shock peening and vibratory finishing and its effect on fatigue strength of Ti-3.5Mo-6.5Al-1.5Zr-0.25Si titanium alloy, J. Alloy. Compd. **783**, 828–835 (2019)

21. L.C. Zhou, Y.H. Li, W.F. He, G.Y. He, X.F. Nie, D.L. Chen, Z.L. Lai, Z.B. An, Deforming TC6 titanium alloys at ultrahigh strain rates during multiple laser shock peening. Mater. Sci. Eng. A-Struct. Mater. Prop. Microstruct. Process. **578**, 181–186 (2013)

22. R. Fabbro, J. Fournier, P. Ballard, D. Devaux, J. Virmont, Physical study of laser-produced plasma in confined geometry. J. Appl. Phys. **68**(2), 775–784 (1990)

23. P. Peyre, C. Carboni, A. Sollier, L. Berthe, C. Richard, E.D.L. Rios, R. Fabbro, New trends in laser shock wave physics and applications. High-power Laser Ablation IV (2002)

24. Y. Yang, K. Zhou, G. Li, Surface gradient microstructural characteristics and evolution mechanism of 2195 aluminum lithium alloy induced by laser shock peening. Opt. Laser Technol. **109**, 1–7 (2019)

25. D. Karthik, K.U. Yazar, A. Bisht, S. Swaroop, C. Srivastava, S. Suwas, Gradient plastic strain accommodation and nanotwinning in multi-pass laser shock peened 321 steel. Appl. Surf. Sci. **487**, 426–432 (2019)

26. B. Mao, Y. Liao, B. Li, Gradient twinning microstructure generated by laser shock peening in an AZ31B magnesium alloy. Appl. Surf. Sci. **457**, 342–351 (2018)

27. Y. Liao, G.J. Cheng, Controlled precipitation by thermal engineered laser shock peening and its effect on dislocation pinning: Multiscale dislocation dynamics simulation and experiments. Acta Mater. **61**(6), 1957–1967 (2013)

28. C. Ye, Y.L. Liao, S. Suslov, D. Lin, G.J. Cheng, Ultrahigh dense and gradient nano-precipitates generated by warm laser shock peening for combination of high strength and ductility. Mater. Sci. Eng. A-Struct. Mater. Prop. Microstruct. Process. **609**, 195–203 (2014)

29. B.P. Fairand, A.H. Clauer, J.F. Ready, Use of laser generated shocks to improve the properties of metals and alloys **86**, 112–121 (1976)

30. I. Altenberger, E.A. Stach, G. Liu, R.K. Nalla, R.O. Ritchie, In situ transmission electron microscope study of the thermal stability of near-surface microstructures induced by deep rolling and laser-shock peening. Scr. Mater. **48**(12), 1593–1598 (2003)

31. V.A. Yanushkevich, Y.N. Nikiforov, M.M. Nishchenko, B.P. Kovalyuk, V.B. Glad'o, V.S. Mocharskii, Effect of improvement of corrosion resistance of 15Kh13MF steel irradiated by laser in shock wave generation mode. Inorg. Mater. Appl. Res. **4**(2), 160–164 (2013)

32. B.Y. Cao, D.H. Lassila, M.S. Schneider, B.K. Kad, C.X. Huang, Y.B. Xu, D.H. Kalantar, B.A. Remington, M.A. Meyers, Effect of shock compression method on the defect substructure in monocrystalline copper. Mater. Sci. Eng. A-Struct. Mater. Prop. Microstruct. Process. **409**(1–2), 270–281 (2005)

33. Y. Fan, Y. Wang, S. Vukelic, Y.L. Yao, Wave-solid interactions in laser-shock-induced deformation processes. J. Appl. Phys. **98**(10) (2005)

34. P. Peralta, D. Swift, E. Loomis, C.H. Lim, K.J. McClellan, Deformation and fracture in laser-shocked NiAl single crystals and bicrystals. Metall. Mater. Trans. A-Phys. Metall. Mater. Sci. **36A**(6), 1459–1469 (2005)

35. B.N. Mordyuk, Y.V. Milman, M.O. Lefimov, G.I. Prokopenko, V.V. Silberschmidt, M.I. Danylenko, A.V. Kotko, Characterization of ultrasonically peened and laser-shock peened surface layers of AISI 321 stainless steel. Surf. Coat. Technol. **202**(19), 4875–4883 (2008)

36. C.H. Ye, S. Suslov, B.J. Kim, E.A. Stach, G.J. Cheng, Fatigue performance improvement in AISI 4140 steel by dynamic strain aging and dynamic precipitation during warm laser shock peening. Acta Mater. **59**(3), 1014–1025 (2011)

37. J.Z. Lu, K.Y. Luo, Y.K. Zhang, G.F. Sun, Y.Y. Gu, J.Z. Zhou, X.D. Ren, X.C. Zhang, L.F. Zhang, K.M. Chen, C.Y. Cui, Y.F. Jiang, A.X. Feng, L. Zhang, Grain refinement mechanism of multiple laser shock processing impacts on ANSI 304 stainless steel. Acta Mater. **58**(16), 5354–5362 (2010)

38. J.Z. Lu, K.Y. Luo, Y.K. Zhang, C.Y. Cui, G.F. Sun, J.Z. Zhou, L. Zhang, J. You, K.M. Chen, J.W. Zhong, Grain refinement of LY2 aluminum alloy induced by ultra-high plastic strain during multiple laser shock processing impacts. Acta Mater. **58**(11), 3984–3994 (2010)

39. X.Z. Chen, J.J. Wang, Y.Y. Fang, B. Madigan, G.F. Xu, J.Z. Zhou, Investigation of microstructures and residual stresses in laser peened Incoloy 800H weldments. Opt. Laser Technol. **57**, 159–164 (2014)

40. G. Chandrasekar, C. Kailasanathan, D.K. Verma, Investigation on un-peened and laser shock peened weldment of Inconel 600 fabricated by ATIG welding process. Mater. Sci. Eng. A-Struct. Mater. Prop. Microstruct. Process. **690**, 405–417 (2017)

41. S.A.N.J. Reddy, S. Prabhakaran, S. Kalainathan, N. Arivazhagan, M. Manikandan, Effect of laser shock peening to improve metallurgical and mechanical properties of Alloy C-276 fabricated by Gas Tungsten Welding Techniques. Laser Eng. **42**(4) (2017)

Chapter 2
Characteristics of Laser-Induced Plasma Shock Wave in Metal Materials

2.1 Introduction

According to the basic principle of LSP, short-pulse high-power laser irradiates materials to form high-pressure plasma to produce shock waves, which act on materials and induce them to produce a gradient stress field and microstructures [1–3]. Shock wave is the energy carrier to strengthen the materials.

Pressure is the most important physical parameter of laser shock wave. At present, the Fabbro's pressure model [4] is often used in research, which is suitable for the Nd:glass laser with the spatial-uniform-distribution spot energy. However, the studies of this book mainly use the YAG solid-state laser, and its spot energy is of Gaussian distribution, and the time–space waveform is quite different from that of the Nd:glass laser. In the experiment, it is found that the measured shock wave pressure triggered by the same power density is usually about twice the calculated value of the Fabbro's pressure model, so the Fabbro's model is not suitable for a LSP process with Gaussian distribution of laser energy space. Therefore, on the basis of analyzing the principle of the formation of the laser shock wave, this book uses PDV (photon doppler velocimeter) to test the shock wave characteristics under different LSP process parameters (such as power density, confinement layer and absorbing layer), and obtains the law of the influence of process parameters on laser shock wave, and then fits the relationship between laser process parameters and shock wave pressure under the YAG laser, and puts forward a new model for the estimation of the pressure of laser shock waves, which can provide guidance for the LSP process of the YAG solid-state laser.

© Zhejiang University Press 2021
L. Zhou and W. He, *Gradient Microstructure in Laser Shock Peened Materials*, Springer Series in Materials Science 314,
https://doi.org/10.1007/978-981-16-1747-8_2

2.2 The Principle of the Formation of the Laser-Induced Plasma Shock Wave

2.2.1 The Process of the Formation of the Laser-Induced Plasma Shock Wave

The interaction between laser and matter (including solid, liquid, vapor and plasma) is a complex process, including the thermal effect of laser on matter, the absorption and reflection of laser by matter and the propagation of laser in the matter [5]. The mechanism of interaction between laser and matter is complex (as shown in Fig. 2.1), including bremsstrahlung absorption, photoionization, multiphoton absorption, hole absorption and impurity absorption [6].

When the high-power laser beam acts on the surface of a solid target, the target surface absorbs a large amount of laser energy, which causes a rise in temperature, melting, gasification, splashing and other phenomena. The specific physical process mainly depends on the laser power density, as shown in Table 2.1.

With the different laser intensities, the mechanism of interaction may be completely different, and the phenomena occurring on the surface of the material

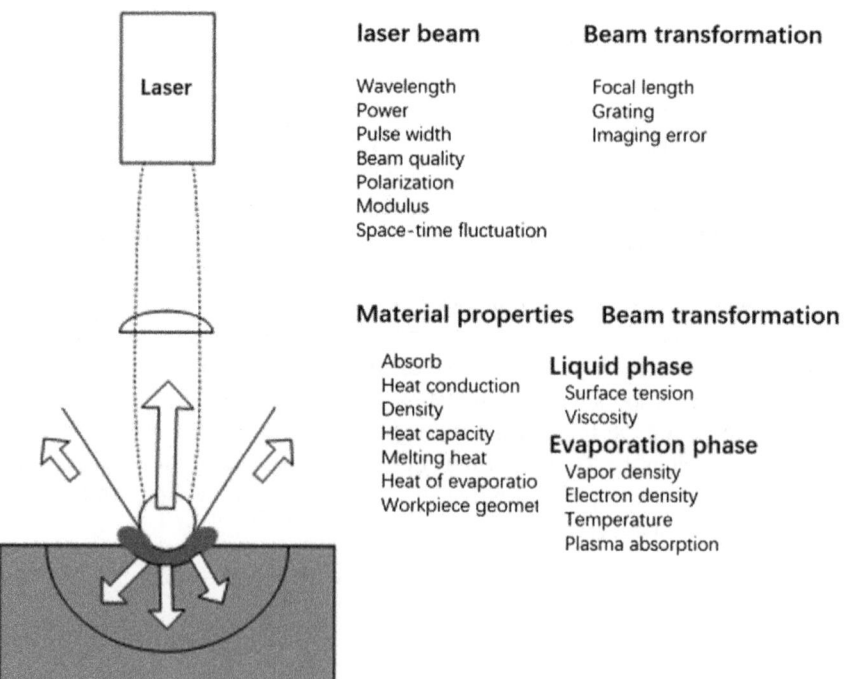

Fig. 2.1 Interaction between laser and matter

Table 2.1 Corresponding relationship between laser power density and physical phenomena

10^3–10^4 W/cm^2	10^4–10^6 W/cm^2	10^6–10^8 W/cm^2	10^8–10^{10} W/cm^2
Heat	Melt	Gasification	Plasma

are also different under different laser power densities. It can be seen from Table 2.1 that the interaction between laser and matter has different physical processes in different power densities. When the laser power density is in the range of 10^3–10^4 W/cm^2, the main thermal effect of the laser on the surface of the material is to heat the surface of the material. The effect of a long pulse laser with power density less than 10^6 W/cm^2 is also a thermal effect, but the material will melt. When the laser power density is greater than 10^6 W/cm^2 and the pulse duration is greater than 100 us, the interaction between laser and material is basically a gasification process. When it exceeds 10^9 W/cm^2, plasma will appear. The LSP process belongs to the last one. When high-power laser irradiates the surface of the material, it directly gasifies and ionizes the absorbing layer and forms high-temperature and high-pressure plasma. The plasma will further absorb laser energy and evolve into shock waves [7, 8].

The process of the formation of a laser-induced plasma shock wave can be divided into three stages [9]: in the first stage, the absorbing layer absorbs energy and melts and gasifies; in the second stage, gas particles absorb energy to form high-temperature and high-pressure plasma; in the third stage, the plasma expands to produce an instantaneous high-pressure shock wave. Specifically, when the surface of metal material is irradiated by a high power density laser, the temperature of aluminum foil will rise rapidly and reach a molten state, and when it reaches a certain degree, it will be gasified into aluminum gas. Aluminum gas continuously absorbs energy through reverse bremsstrahlung absorption and photoionization, and finally forms high-temperature and high-pressure plasma. Plasma absorbs laser energy through normal absorption (inverse bremsstrahlung absorption) and abnormal absorption (collision absorption), which increases the degree and temperature of its ionization. With the continuous expansion of plasma after absorbing energy, a high-pressure shock wave will be generated instantly, and the pressure of the shock wave can reach a GPa level under the action of the water confinement layer. According to whether the shock wave propagates at subsonic velocity or supersonic velocity relative to gas (gas and ambient air), it can be divided into two categories: aLSA wave propagating at subsonic velocity is called a Laser Supported Combustion Wave (LSCW/LSC); a LSA wave propagating at supersonic velocity is a Laser Supported Detonation Wave (LSDW/LSD) [10–12] (as shown in Fig. 2.2).

An LSC or LSD phenomenon occurs corresponding to different ranges of laser intensity. When the gasification on the target surface is strong, the gas of the target is partially ionized and heated, and then the cold air in front is heated and ionized by heat radiation, forming an LSC wave. At this time, some of the laser still irradiates on the target surface through the plasma area. The radiation of plasma near the target helps to enhance the thermal coupling between laser and target, which weakens with the departure of plasma and gradually forms a shield for the target. With the

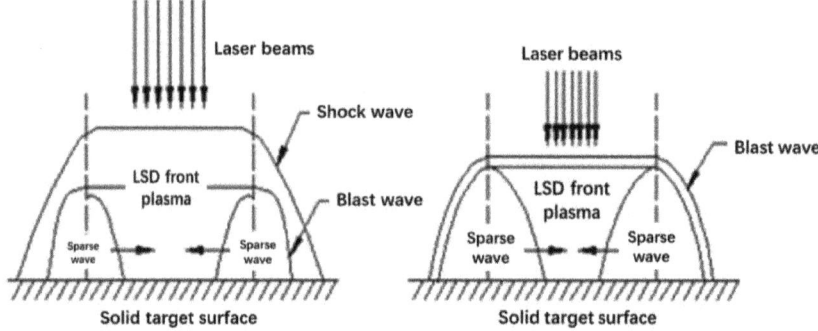

(a) Laser supported combustion wave LSC   (b) Laser supported detonation wave LSD

Fig. 2.2 Structure diagram of LSC and LSD

increase in laser intensity, the LSC absorption area moves faster, the absorption is strengthened, and it converges with the front shock wave to form an LSD wave, which constitutes the complete absorption of laser. The main influence of the LSD wave on the target surface is the increase in the pressure of the flow field after the wave, that is, the impulse coupling between laser and target is enhanced. As the rarefaction waves reach the central spot, the pressure of the flow field attenuates and the impulse coupling is weakened.

2.2.2 The Propagation and Attenuation of the Laser-Induced Plasma Shock Wave

After the laser shock wave acts on the material, it will propagate inwards in the form of a stress wave, which mainly includes three types of elastic waves: a surface wave (Rayleigh wave), a longitudinal wave (compression/tension wave) and a transverse wave (shear wave) [13], as shown in Fig. 2.3. "Surface wave" (Rayleigh wave) refers to the stress wave formed by the surface material moving in up-down and back–forth directions under the force of the impact. "Longitudinal wave" (compression/tension wave) refers to the stress wave formed by internal material moving back and forth along the direction of the stress, and the direction of the movement of its material particles is the same as the direction of the propagation of the stress wave. "Transverse wave" (shear wave) refers to the stress wave formed by internal material moving perpendicular to the direction of the propagation. Among them, the surface wave is the slowest wave, while the longitudinal wave is the fastest one. In the process of LSP, the laser-induced shock wave will act vertically on the surface of the material, thus forming a stress wave and propagating inward, and the stress wave will also propagate in the form of a surface wave, a compression wave and a shear wave,

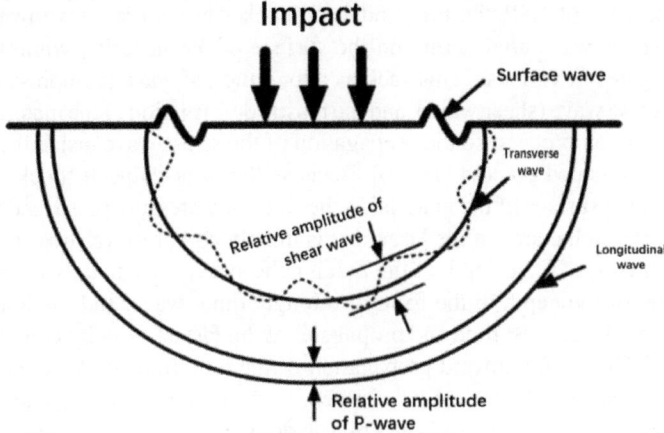

Fig. 2.3 The stress wave system under surface impact

among which the compressive wave (longitudinal wave) in the depth direction is the main form [14].

According to the shock wave compression theory of solids [15], three different mechanical responses of materials will occur under the action of a shock wave (see Fig. 2.4): (1) strong impact; (2) elastic impact; (3) elastic–plastic impact. "Strong impact" refers to the drastic change in the state of materials under a narrow shock wave front (pressure above tens of GPas); the material only undergoes elastic deformation under an elastic impact. The elastic–plastic impact is between them, and the pressure of the shock wave on the material exceeds the dynamic elastic limit of the material, resulting in plastic deformation. According to the theoretical model of the peak pressure of a laser-induced shock wave and the actual test results, it can be seen that the pressure can reach several GPas, but the rising edge of the shock wave is not very steep, which belongs to medium-strength impart, so the material will show the mechanical response of elastic–plastic impart [16, 17].

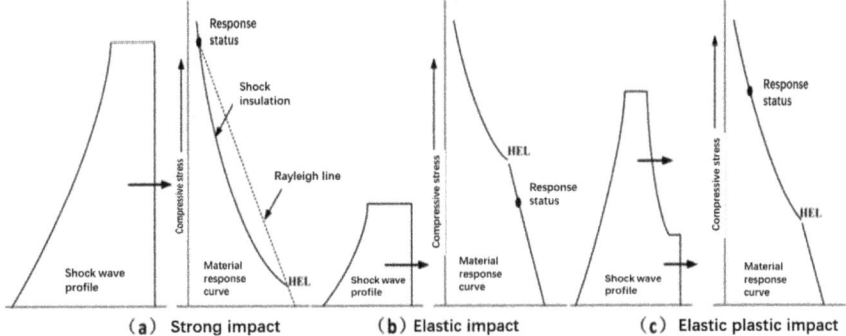

Fig. 2.4 The characteristic curve of the dynamic response of material under a shock wave

In the process of LSP, the laser-induced shock wave propagates inward in the form of a stress wave after acting on the surface of the material, while the stress wave propagates in various forms such as a longitudinal wave (compression wave) and a transverse wave (shear wave), and the plastic deformation of the material occurs in the propagation process, so the propagation of the stress wave inside the material is a very complicated process [18–20]. Because the laser-induced shock wave acts vertically on the surface of the material in the form of a pressure pulse, and the stress on the material in the area of the laser spot is mainly compressive stress in the axial direction (direction of the depth of the Z axis), the stress wave mainly propagates in the direction of the depth in the form of a longitudinal wave, and the longitudinal wave forms in the earliest time and propagates at the fastest velocity compared with other waves. During the inward propagation of the longitudinal wave, the material will be compressed and plastically deformed, so the propagation of the longitudinal wave also exists in the form of elastic–plastic double waves. At the initial stage, the shock wave propagates into the material in the form of a stress wave. Because the axial compressive stress does not reach the dynamic elastic limit of the material, the material will not be plastically deformed, and the stress wave presents a single wave structure of an elastic wave. With the pressure of the shock wave rising, plastic deformation of the material begins to occur. With the generation of plastic strain, the elastic wave single-wave structure of the stress wave is broken, and the double-wave structure of the plastic wave and the elastic wave propagates in the direction of the depth. When the shock wave is loaded to the maximum, the plastic deformation of the material reaches its maximum. With the unloading of shock wave pressure and the inward propagation of the maximum pressure wave front, the plastic deformation of the surface of the material has basically completed, and the free surface velocity of the back reaches its maximum when the maximum pressure wave reaches the back surface. In this process, the stress wave propagates inward and plastic deformation occurs, the pressure of the stress wave is attenuated, the plastic wave is unloaded, and the stress wave continues to propagate inward in the form of an elastic wave in the later stage.

Therefore, the path of the first wave of the inward propagation of the stress wave during LSP can be divided into three propagation stages: elastic wave, elastic–plastic double wave and elastic wave [21–23], and the plastic deformation inside the material is basically completed during the path of the first wave of the stress wave. Under the influence of plastic deformation, damping consumption and energy conversion, the stress wave will be attenuated in the process of inward propagation, resulting in the gradual reduction in the pressure of the stress wave. During the stage of elastic–plastic double wave propagation, a large amount of plastic deformation will lead to the accelerated attenuation of the stress wave.

2.3 The Characteristics of Laser-Induced Plasma Shock Waves Under Different Process Parameters

2.3.1 Test Method For the Characteristics of Laser-Induced Plasma Shocks

In this chapter, PDV (as shown in Figs. 2.5 and 2.6) is mainly used to test the characteristics of the shock wave, which has the advantages of non-contact, a high spatial resolution, a fast dynamic response, a wide measuring range and high directional sensitivity. The test principle is as follows:

When the light emitted by the light source irradiates the moving object, the probe receives the light reflected by the object. Compared with the frequency f_0 of the incident light (signal light), the reflected light frequency f_s will have a frequency shift, which is called the Doppler frequency shift f_d [24]:

$$f_d = f_s - f_0 = \frac{2V}{\lambda_0} = \frac{2f_0V}{c} \tag{2.1}$$

where f_s is the frequency of the probe light (reflected light), f_0 is the frequency of the signal light (incident light source), V is the velocity at which the object moves towards the probe, λ_0 is the central wavelength of the probe light, and c is the velocity of light in the vacuum.

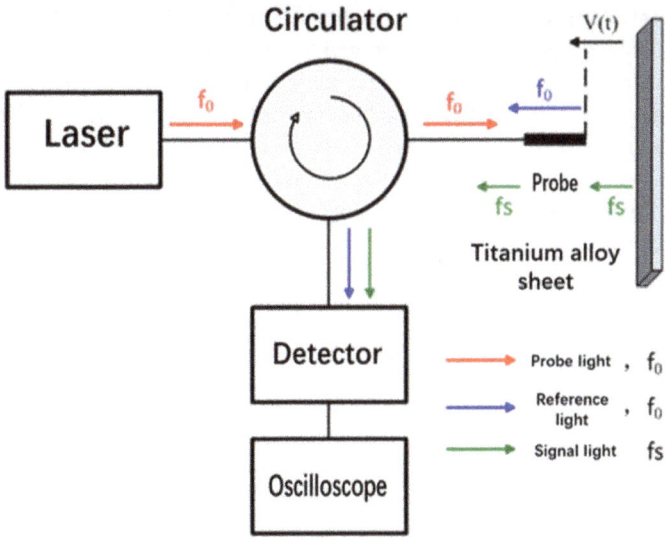

Fig. 2.5 Schematic diagram of the testing principle of the photon doppler velocimeter

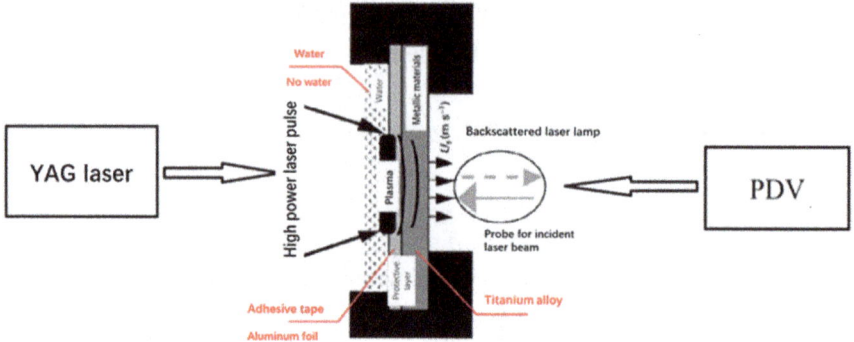

Fig. 2.6 PDV test method of a laser-induced shock wave

The laser emits a signal light with a sufficient density of power. After passing through the circulator, the signal light is divided into two paths. One path enters the probe through the circulator, and the probe emits a light source with frequency f_0. The light source irradiates the titanium alloy sheet with a moving velocity V(t) (i.e. the velocity of the free surface on the back of the sheet after the laser strikes the sheet). At this time, the light source is reflected on the sheet surface. According to the photon Doppler effect, the frequency of the reflected light becomes light with a Doppler frequency shift, and the frequency is f_s. Part of the reflected light will enter the probe and return to the circulator along the original path, and then reach the detector. The reflected light at this time is called the probe light. Another light source directly enters the detector through the circulator, and this light source is called the reference light at this time, and the frequency is still f_0. The probe light interferes with the reference light in the detector, and the interference light is converted into a voltage signal, and the output voltage signal is collected by oscilloscope. In the test process, because the detector and oscilloscope will produce a certain noise, and the noise frequency is usually 0–100 MHz, which is very close to the frequency of the target signal. Therefore, the short Fourier transform is used to process the data, and the velocity history of the moving target can be obtained.

In the process of LSP, the main factor to improve the surface properties of materials is the pressure of the shock wave. When the laser irradiates the target surface, plasma is generated, and a shock wave pressure P plasma is generated after the plasma explodes. We approximately deem that the shock wave $\sum_{target}$ applied to the surface of the material under the constraint condition is the same as the pressure of the shock wave generated by the plasma, that is, $\sum target = P_{plasma}$. Assuming that the impact and release parts of the Hugoniot curve are almost symmetrical (completely symmetrical when the stress level is (2HEL)), the measured free surface velocity Uf is about twice the area particle velocity U before impact ($U_f = 2U$).

when $\sum < HEL$,

$$\sum = \rho_0 C_{el} U \tag{2.2}$$

when $\sum > HEL$,

$$\sum = \rho_0 DU + 2/3\sigma_Y = \rho_0(C_0 + SU)U + 2/3\sigma_Y \qquad (2.3)$$

where $\sum$ is the peak pressure generated by the laser-induced shock wave, ρ_0 is the density of the TC4 titanium alloy sheet, C_0 is the velocity of the longitudinal elastic wave of the titanium alloy sheet, U is the velocity of the particles in the impact area, $2/3\sigma_Y$ is the elastic contribution to the deviated part in the process of shock wave generation, and σ_Y is the strength of the dynamic yield.

This chapter is mainly about testing the characteristics of the shock wave induced by a YAG solid-state laser. The models and parameters used are shown in Table 2.2. The measured waveform is shown in Fig. 2.7.

The material used in the test is aviation material TC4 alloy, which belongs to the α- and β-phase alloy with medium strength, with the stabilizing elements 96%

Table 2.2 Parameters of sgr-60 laser

Model	Laser wavelength	Laser pulse width	Spot diameter	Output energy	Working frequency
SGR-60	1064 nm	20 ns	2.6 mm	2–5.8 J	1 Hz Single trigger

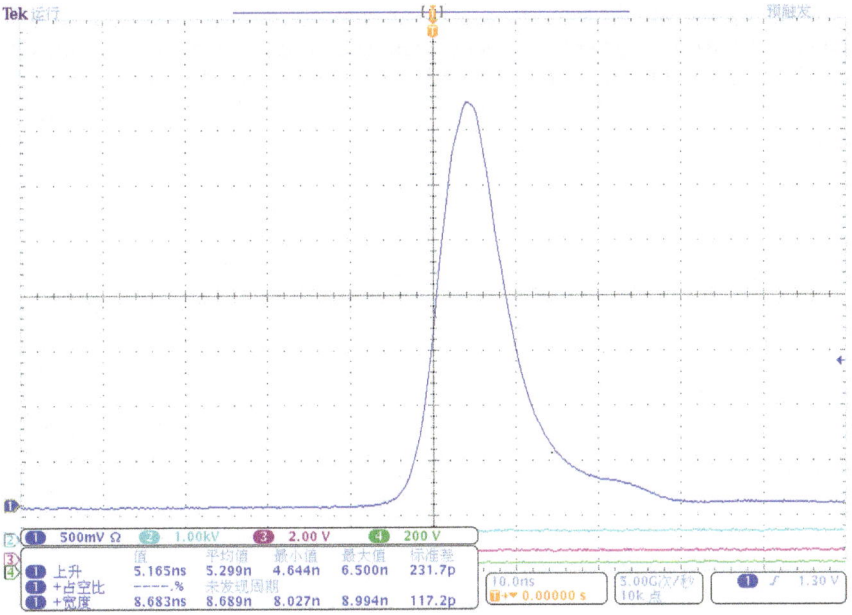

Fig. 2.7 Measured waveform of the laser output pulse

Fig. 2.8 TC4 sheet to be tested

α-phase and 4% β-phase. The size of the TC4 titanium alloy sheet is 35 mm × 35 mm × 1.2 mm, as shown in Fig. 2.8.

The laser shock wave PDV test platform, as shown in Fig. 2.9, clamps the PDV probe on the height bracket and puts it on a stable experimental platform. Open the laser light path, and directly irradiate the light spot on the probe, so that the light path and the probe are always on the same axis. The reason to do so is that the deformation of the material is arc-shaped during LSP, and the dynamic response of the position of the maximum deformation (i.e., the center position) can be accurately measured by keeping the same axis. A synchronizer is connected between the LSP equipment and the Doppler velocimeter equipment, and the signal is transmitted to the oscilloscope at the instant of impact, so that the time of LSP on the surface of the target is synchronized with the PDV test response time to ensure the accuracy of the test.

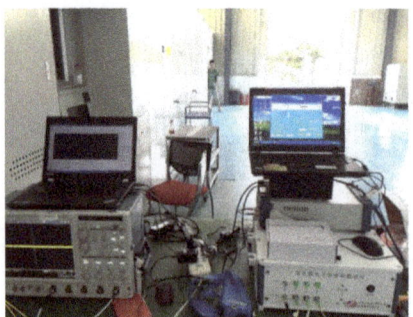 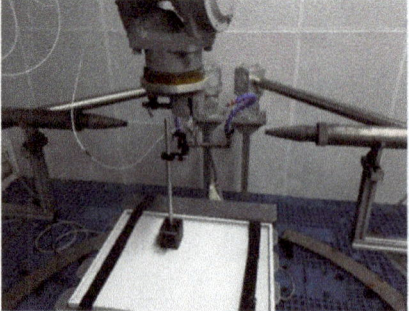

(a) Photon Doppler velocimeter and oscilloscope (b) Adjusting the height of the optical fiber probe

Fig. 2.9 PDV test platform

2.3.2   Laser Power Density

(1)   The influence of laser power density when the absorbing layer is adhesive tape

The adhesive tape serves as the absorbing layer, and the target material is TC4 with a thickness of 500 μm. The velocity of the free surface particles measured under different laser power density conditions is shown in Fig. 2.10. When the laser power density is 4.71 GW/cm^2, the velocity of the free surface particles reaches the maximum value of 736.32 m/s. When the laser power density is less than 4.71 GW/cm^2, the higher the laser power density, the higher the velocity of the free surface particles velocity. When the laser power density is greater than 4.71 GW/cm^2, the velocity of the free surface particles decreases.

(2)   The influence of laser power density when the absorbing layer is aluminum foil

The absorbing layer is made of aluminum foil, and the thickness of the target is TC4 with a thickness of 500 μm. Under different conditions of laser power density (1.88, 2.82, 3.77, 4.71, 5.46 GW/cm^2 in turn), the measured velocity of the free surface particles is shown in Fig. 2.11. When the laser power density is 4.71 GW/cm^2, the velocity of the free surface particles reaches the maximum value of 667.21 m/s. When the laser power density is less than 4.71 GW/cm^2, the higher the laser power density, the higher the velocity of the free surface particles. When laser power density is greater than 4.71 GW/cm^2, the higher the laser power density, the lower the velocity of the free surface particles.

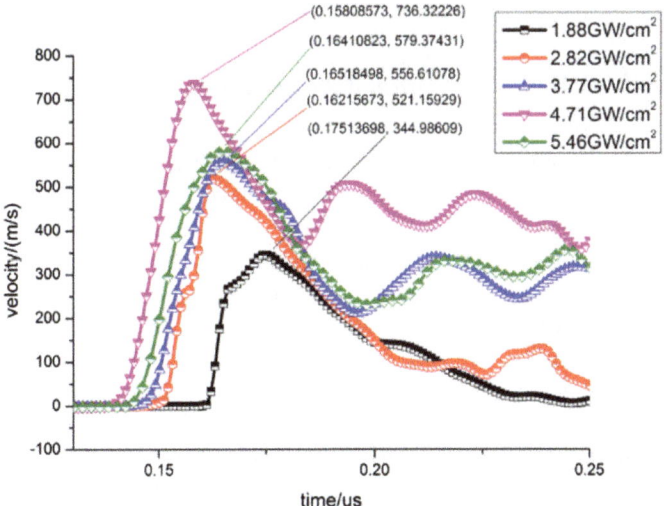

Fig. 2.10 The law of change in the velocity of particles on a TC4 free surface subjected to LSP (absorbing layer: adhesive tape)

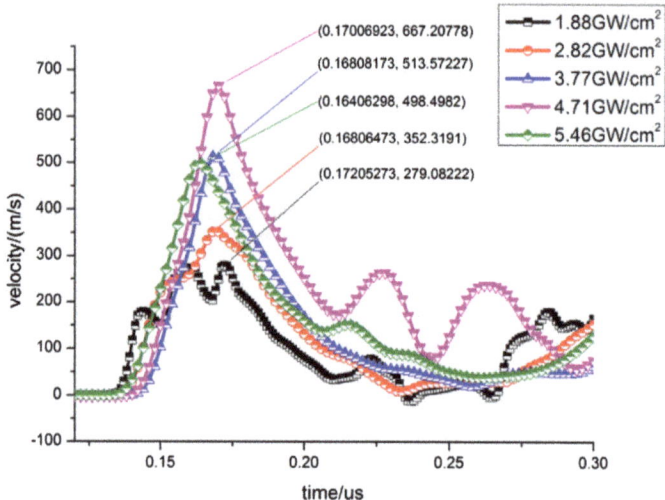

Fig. 2.11 Variation of particle velocity on a TC4 free surface subjected to LSP (absorbing layer: aluminum foil)

(3)   The influence of laser power density without an absorbing layer

Without an absorbing layer, the target material is TC4 with a thickness of 500 μm. Under different laser power densities (2.82, 3.77, 4.71 GW/cm^2 in turn), the measured velocity of the free surface particles is shown in Fig. 2.12. With the increase in laser power density, the velocity of the free surface particles increases.

2.3.3 With or Without a Water Confinement Layer

(1)   When the absorbing layer is adhesive tape, the difference between when there is and when there isn't a water confinement layer

The laser power density is 4.71 GW/cm^2, the absorbing layer is adhesive tape, and the target material is TC4 with a thickness of 500 μm. LSP is carried out under the conditions of a water confinement layer and no water confinement layer, and the measured velocity of the free surface particles is shown in Fig. 2.13. When there is a water confinement layer in the LSP process, the maximum velocity of the free surface reaches 736.32 m/s, which keeps a high velocity within 1 μs; the maximum velocity of the free surface is only 41.85 m/s when there is no confinement layer during LSP. The water confinement layer can effectively improve the velocity of the free surface particles when the target is subjected to LSP.

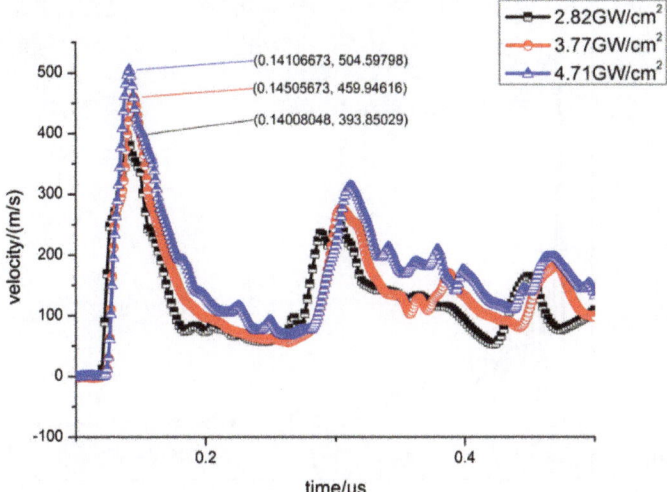

Fig. 2.12 The variation of particle velocity on a TC4 free surface subjected to LSP without an absorbing layer

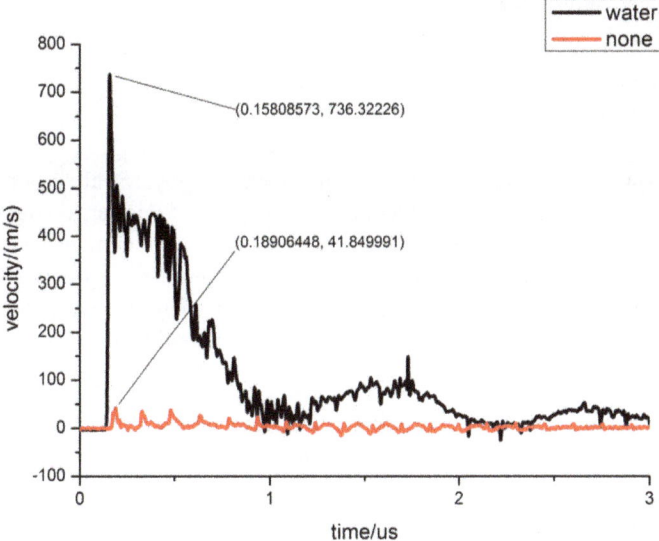

Fig. 2.13 When the absorbing layer is adhesive tape, the influence of a water confinement layer on the velocity of the particles of the free surface

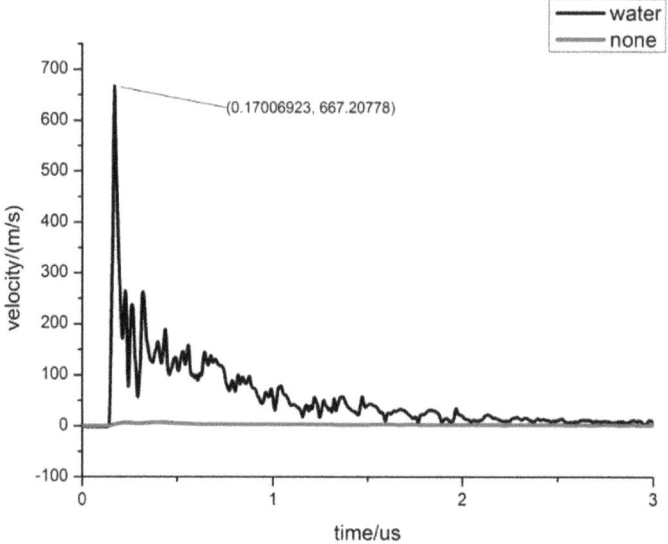

Fig. 2.14 When the absorbing layer is aluminum foil, the influence of the water confinement layer on the velocity of the particles of free surface

(2)   When the absorbing layer is aluminum foil, the difference between when there is and when there isn't a water confinement layer

The laser power density is 4.71 GW/cm², the absorbing layer is aluminum foil, and the target material is TC4 with a thickness of 500 μm. LSP is carried out under the conditions of a water confinement layer and no water confinement layer, and the measured velocity of the free surface particles is shown in Fig. 2.14. The maximum velocity of the free surface reaches 667.21 m/s when there is a water confinement layer in the LSP, and only 7.10 m/s when there is no confinement layer in the LSP process.

(3)   When there is no absorbing layer, the difference between when there is and when there isn't a water confinement layer

The laser power density is 4.71 GW/cm², there is no absorbing layer, and the target material is TC4 with a thickness of 500 μm. LSP is carried out under the conditions of a water confinement layer and of no water confinement layer, and the measured velocity of the free surface particles is shown in Fig. 2.15. The maximum velocity of free surface reaches 504.60 m/s when there is a water confinement layer in the LSP process, and only 30.89 m/s when there is no confinement layer in the LSP process.

The relationship between the calculated peak pressure of free surface shock wave and the laser power density is shown in Fig. 2.16. With an increase in the laser power density, the peak pressure of the shock wave gradually rises, reaching a maximum value of 9.71 GPa at 4.71 GW/cm², and the peak pressure of the shock wave decreases

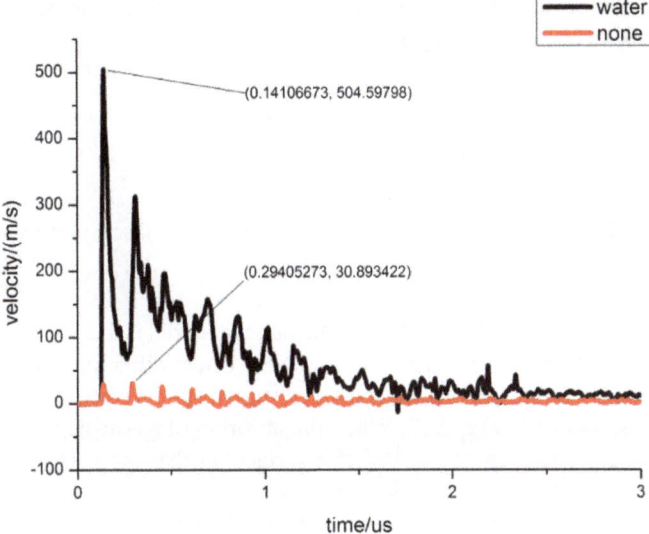

Fig. 2.15 When there is no absorbing layer, the influence of a water confinement layer on the velocity of the particles on the free surface

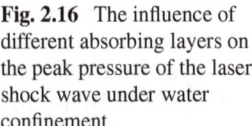

Fig. 2.16 The influence of different absorbing layers on the peak pressure of the laser shock wave under water confinement

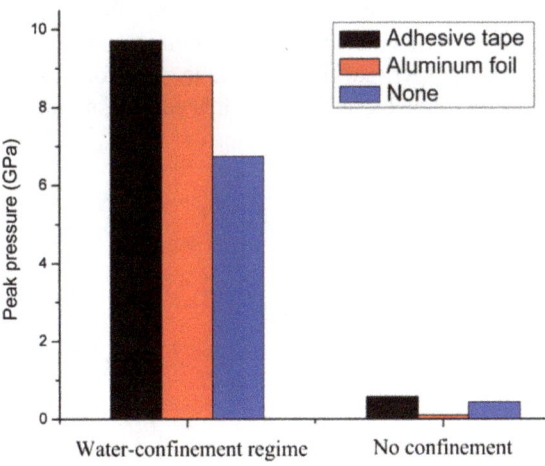

with an increase in the laser power density. The reason for this phenomenon is that when the laser power density reaches a high value, the water confinement layer is broken down, so the laser energy is lost, and the pressure of the shock wave is reduced, thus the velocity of the particles on the free surface of the target is reduced.

It can be seen from the histogram (Fig. 2.16) that the water confinement layer has a great influence on laser-induced shock waves, and the pressure of the shock wave under the confinement layer can be increased by 10–50 times compared with that when there is no confinement layer.

2.3.4 Absorbing Layer

(1) The influence of the absorbing layer when the laser power density is 1.88 GW/cm^2

The laser power density is 1.88 GW/cm^2, the target material is TC4 with a thickness of 500 μm, and there is a water confinement layer. Under different absorbing layers (adhesive tape or aluminum foil), the measured velocity of the particles on the LSPed free surface is shown in Fig. 2.17. When the absorbing layer is adhesive tape, the velocity of the particles on the LSPed free surface reaches the maximum value of 364.87 m/s, and when the absorbing layer is aluminum foil, the velocity of the particles on the LSPed free surface reaches the maximum value of 273.09 m/s.

(2) The influence of the absorbing layer whenthe laser power density is 2.82 GW/cm^2

The laser power density is 2.82 GW/cm^2, the target material is TC4 with a thickness of 500 μm, and there is a water confinement layer. Under different conditions of the absorbing layer (adhesive tape, aluminum foil or no absorbing layer), the measured velocity of the particles on the LSPed free surface is shown in Fig. 2.18. When the

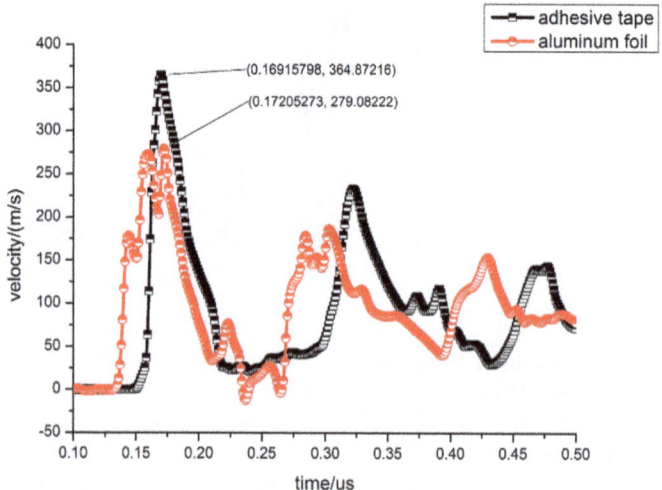

Fig. 2.17 When laser power density is 1.88 GW/cm^2, the relationship between the different absorbing layers and the velocity of the particles on the free surface

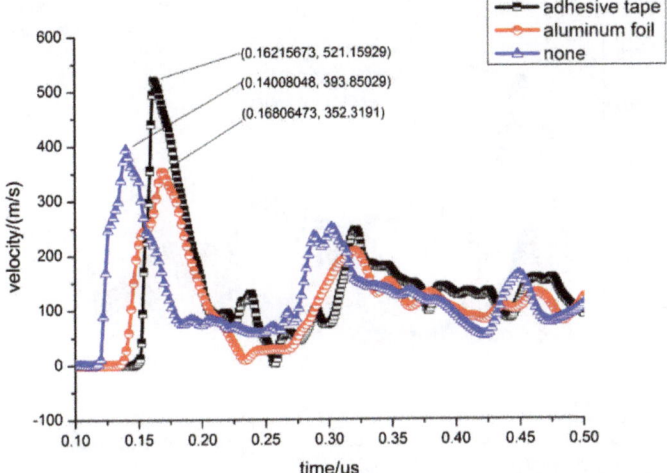

Fig. 2.18 When laser power density is 2.82 GW/cm², the relationship between the different absorbing layers and the velocity of the particles on the free surface

absorbing layer is adhesive tape, the maximum velocity of the particles on the LSPed free surface is 521.16 m/s, when the absorbing layer is aluminum foil, the maximum velocity of the particles on the LSPed free surface is 352.32 m/s, and when there is no absorbing layer, the maximum velocity of the particles on the LSPed free surface is 393.85 m/s.

(3)   The influence of the absorbing layer whenthe laser power density is 3.77 GW/cm²

Laser power density is 3.77 GW/cm², the target material is TC4 with a thickness of 500 μm, and there is a water confinement layer. Under different conditions of the absorbing layer (adhesive tape, aluminum foil or no absorbing layer), the measured velocity of the particles on the laser-impacted free surface is shown in Fig. 2.19. When the absorbing layer is adhesive tape, the maximum velocity of the particles on the laser-impacted free surface is 556.61 m/s, when the absorbing layer is aluminum foil, the maximum velocity of the particles on the laser-impacted free surface is 513.57 m/s, and when there is no absorbing layer, the maximum velocity of the particles on the laser-impacted free surface is 459.95 m/s.

(4)   The influence of the absorbing layer whenthe laser power density is 4.71 GW/cm²

The laser power density is 4.71 GW/cm², the target material is TC4 with a thickness of 500 μm, and there is a water confinement layer. Under different conditions of the absorbing layer (adhesive tape, aluminum foil or no absorbing layer), the measured velocity of the particles on the LSPed free surface is shown in Fig. 2.20. When the

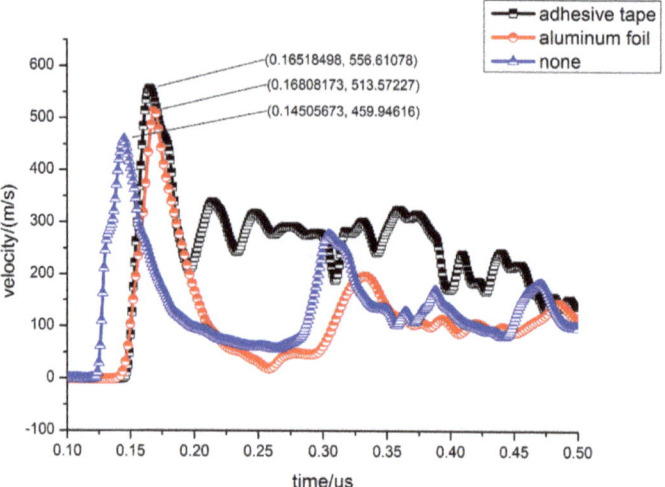

Fig. 2.19 When laser power density is 3.77 GW/cm^2, the relationship between the different absorbing layers and the velocity of the particles on the free surface

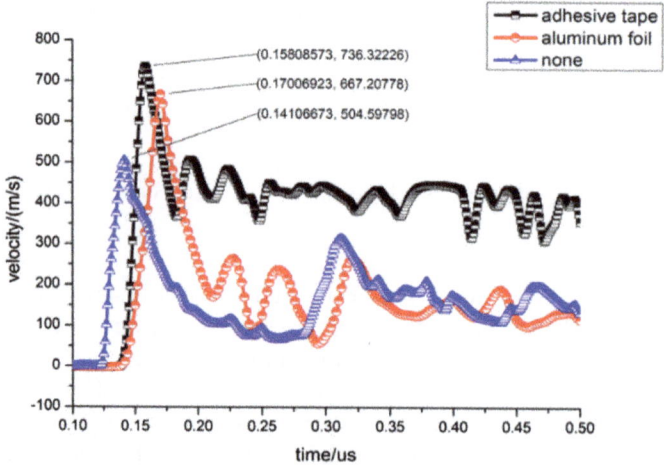

Fig. 2.20 When laser power density is 4.71 GW/cm^2, the relationship between the different absorbing layers and the velocity of the particles on the free surface

absorbing layer is adhesive tape, the maximum velocity of the particles on the LSPed free surface is 736.32 m/s, when the absorbing layer is aluminum foil, the maximum velocity of the particles on the LSPed free surface is 667.21 m/s, and when there is no absorbing layer, the maximum velocity of the particles on the LSPed free surface is 504.60 m/s.

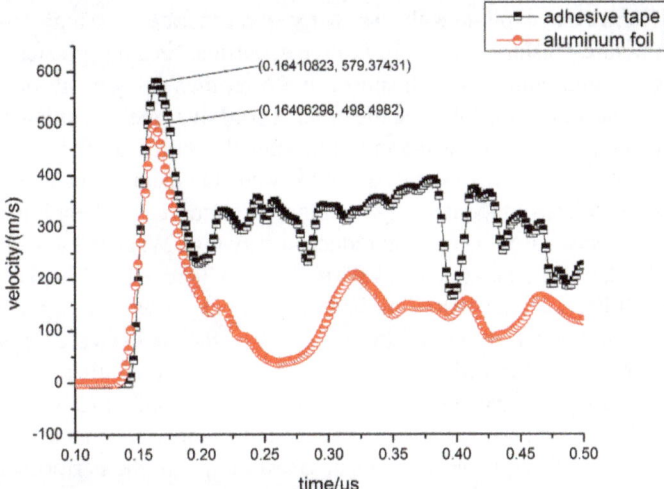

Fig. 2.21   When laser power density is 5.46 GW/cm^2, the velocity changes in different absorbing layers and free surface particles are observed

(5)   The influence of the absorbing layer whenthe laser power density is 5.46 GW/cm^2

The laser power density is 5.46 GW/cm^2, the target material is TC4 with a thickness of 500 μm, and there is a water confinement layer. Under different absorbing layers (adhesive tape or aluminum foil), the measured velocity of the particles on the LSPed free surface is shown in Fig. 2.21. When the absorbing layer is adhesive tape, the velocity of the particles on the LSPed free surface reaches the maximum value of 579.37 m/s, and when the absorbing layer is aluminum foil, the velocity of the particles on the LSPed free surface reaches the maximum value of 498.50 m/s.

The peak pressure of the laser shock wave produced by the adhesive tape as the absorbing layer impacting the target at different densities of laser power is the largest. When laser power density is less than 2.82 GW/cm^2, the peak pressure of the laser shock wave produced without an absorbing layer is greater than that when the absorbing layer is aluminum foil. When laser power density is greater than 2.82 GW/cm^2, the peak pressure of the laser shock wave without an absorbing layer is less than that when the absorbing layer is aluminum foil.

In the process of laser irradiation, a layer of metal on the target surface can be regarded as an absorbing layer. When there is no absorbing layer, because the absorbing layer and the target are the same material and have the same acoustic impedance, the response time of the shock wave will be faster. When adhesive tape and aluminum foil are used as an absorbing layer, the peak pressure of the laser shock wave increases noticeably, and the peak pressure of the laser shock wave under an adhesive tape absorbing layer is higher. Under the same parameters, because the gasification heat of adhesive tape is lower than that of aluminum foil, adhesive tape can be gasified, ionize and explode with less energy and generate more high-temperature and high-pressure plasma. And there is higher plasma, creating greater efficiency when using the aluminum foil as the absorbing layer than using no absorbing layer, so in terms of the quantity of plasma, the order is adhesive tape > aluminum foil > no absorbing layer. The greater the amount of plasma, the more energy is absorbed and the higher the peak pressure of the laser shock wave. On the other hand, the thermal conductivity of adhesive tape is lower than that of aluminum foil, so it is easier for adhesive tape to save laser energy and reduce energy loss. When the absorbing layer exists, the shock wave generated by laser irradiation will be reflected and transmitted back and forth between the absorbing layer and the target, which will increase the action time of the shock wave and the impulses of the shock wave acting on the target, thus improving the LSP effect. When the laser is irradiating on the surface of the absorbing layer, it gasifies to form plasma, which continuously absorbs laser energy, and then it expands rapidly to produce a high-pressure shock wave, which acts on the surface of the material and can greatly improve the performance of the material [25, 26].

2.4   Model of Laser-Induced Plasma Shock Pressure

2.4.1   The Fabbro's Model

Peyer and Fabbro put forward a complete ablation model under a constraint condition in their literature [6, 13], which can work out the final pressure caused by plasma expansion under heating and adiabatic cooling conditions. The laser used in this model is the Nd:glass laser with Gaussian energy-time distribution and uniform spatial distribution. In the model, considering that plasma is an ideal gas, the peak pressure of the shock wave is estimated [27–29]:

$$P(\text{GPa}) = 0.01\sqrt{\frac{\alpha}{2\alpha+3}}\sqrt{Z(\text{gcm}^{-2}\text{s}^{-1})}\sqrt{I_0(\text{GW/cm}^2)} \tag{2.4}$$

In Formula 2.4, I_0 is the power density of the incident laser, P is the peak pressure of the shock wave, α is the efficiency of interaction between laser and substance (αE is the gain of pressure increase, $(1-\alpha)E$ is for generating and ionizing plasma), and Z is the equivalent impedance between target and confinement medium:

$$\frac{1}{Z} = \frac{1}{Z_{靶材}} + \frac{1}{Z_{约束介质}} \tag{2.5}$$

The models commonly used in industrial production are all models under a water confinement layer, which are given by Formula 2.6 according to common parameters ($\alpha = 0.12$, $Z_{\text{water}} = 0.165 \times 10^6$ gcm^{-2} s^{-1}, $Z_{\text{Al}} = 1.5 \times 10^6$ gcm^{-2} s^{-1}):

$$P(\text{GPa}) = 1.02\sqrt{I_0(\text{GW/cm}^2)} \tag{2.6}$$

Formula 2.6 is applicable to the calculation of the peak pressure of the shock wave when the laser pulse energy is uniformly distributed in space and the distribution of the laser pulse time is Gaussian distribution. When the laser power density is greater than 4 GW/cm^2, the peak pressure of the shock wave begins to decrease gradually due to the breakdown of the water confinement layer.

2.4.2 Modified Model

According to the previous results, the adhesive tape absorbing layer and the water confinement layer are selected, and the test parameters under different laser power densities are used to fit the test data. The fitting results are shown in Fig. 2.22. The results of several fittings are as follows [30, 31]:
Linear fitting result:

$$P_{\text{linear\_fit}} = 1.62I_0 + 1.86 \tag{2.7}$$

Polynomial fitting result:

$$P_{\text{polynomial\_fit}} = 0.045I_0^2 + 1.32I_0 + 2.3 \tag{2.8}$$

Nonlinear fitting result 1:

$$P_{nonlinear\_fit\_1} = 3.05I_0^{0.73} \tag{2.9}$$

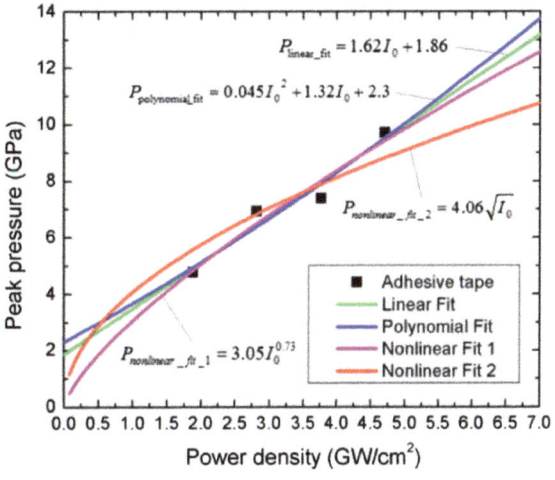

Fig. 2.22 Fitting results of the peak pressure of the shock wave under a protective layer absorbed by a belt

Nonlinear fitting result 2:

$$P_{nonlinear\_fit\_2} = 4.06\sqrt{I_0} \tag{2.10}$$

Comparing the four fitting results, nonlinear fitting result 2 is more in line with the law that the peak pressure of the shock wave changes with the laser power density, so the model of the pressure of the shock wave obtained by fitting under the adhesive tape absorbing layer is as follows:

$$P_{adhesive} = 4.06\sqrt{I_0} \tag{2.11}$$

Tests show that the distribution of the spatial energy has a great influence on the pressure of the laser shock wave. According to Fabbro's model of the pressure of the shock wave (Formula 2.4), when the target material is TC4 titanium alloy, the confinement layer is water and the absorbing layer is aluminum foil, the efficiency of absorption α is taken as 0.25, and Z is the referring impedance of water, aluminum foil and the target material, where $Z_{water} = 0.165 \times 10^6$ gcm^{-2} s^{-1} and $Z_{Al} = 1.5 \times 10^6$ gcm^{-2} s^{-1}, $Z_{TC4} = 2.2 \times 10^6$ gcm^{-2} s^{-1}. The calculated referring impedance is:

$$Z = \frac{3}{\frac{1}{Z_{water}} + \frac{1}{Z_{Al}} + \frac{1}{Z_{TC4}}} = 0.418 \times 10^6 \text{ gcm}^{-2} \text{ s}^{-1} \tag{2.12}$$

Therefore, the model of the pressure of the shock wave of Fabbro can be expressed as:

$$P_{fabbro} = 1.73\sqrt{I_0} \tag{2.13}$$

Comparing the pressure model $P_{aluminum} = 3.49\sqrt{I_0}$ obtained through the fitting under the condition of choosing the aluminum foil as the absorbing layer in the shock wave test, it is found that the peak pressure of the shock wave obtained through fitting in the test is about twice as high as that calculated by the Fabbro's model under the same laser power density. As shown in Fig. 2.23, it can be seen from the literature that the measured peak pressure is about twice as high as the peak pressure of LSP shock waves with the uniform spatial distribution abroad.

Different peak pressures of laser shock waves are caused by different distributions of laser energy. Figure 2.24a is the cross section of the laser generated by Nd:glass with a uniform distribution of spatial pressure obtained by the foreign researchers. Figure 2.24b is the cross section of the laser generated by Nd:YAG with Gaussian distribution of spatial pressure obtained by Chinese researchers.

The volume function of two kinds of spatial distribution is calculated by integrating the solid of revolution in the shaded part of the figure. For the uniform spatial distribution:

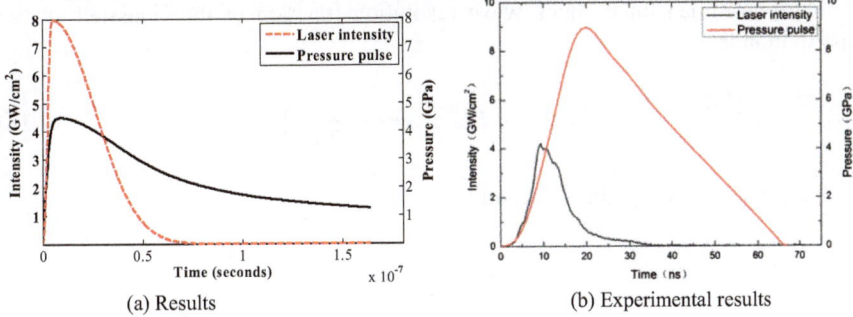

(a) Results   (b) Experimental results

Fig. 2.23   Laser time pulse waveform and shock wave waveform

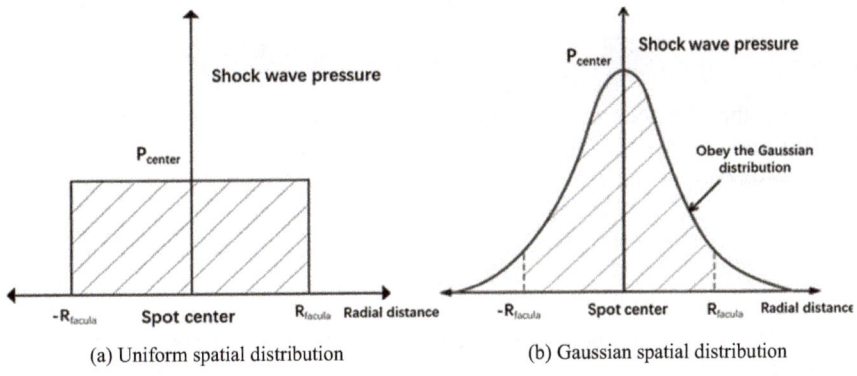

(a) Uniform spatial distribution   (b) Gaussian spatial distribution

Fig. 2.24   A cross section of the distribution of shock wave pressure

$$V_1 \int_0^{P_{center\_1}} \pi R_s^2 \mathrm{dp} \tag{2.14}$$

where P_{center_1} is the center pressure for the uniform spatial distribution and R_s is the laser spot radius 1.3 mm. It can be calculated that:

$$V_1 = 5.31 P_{center\_1} \tag{2.15}$$

For lasers with a uniform spatial distribution, the central pressure of the shock wave is the peak pressure:

$$P_{peak\_ceiling} = P_{center\_1} \tag{2.16}$$

$$V_1 = 5.31 P_{peak\_ceiling} \tag{2.17}$$

The two-dimensional shock wave resolution function of the Gaussian spatial distribution is:

$$P(r) = P_{center\_2} \exp\left(\frac{r^2}{2R^2}\right) \tag{2.18}$$

For the Gaussian spatial distribution:

$$V_1 \int \pi P^2 dp \tag{2.19}$$

$$V_2 = \int\limits_{\exp(-\frac{R_s^2}{2R^2})}^{P_{center\_2}} -2\pi R^2\left(\ln \frac{p}{P_{center\_2}}\right)dp + \int\limits_{0}^{\exp(-\frac{R_s^2}{2R^2})} \pi R_s^2 dp \tag{2.20}$$

where r is the distance from the center of the spot, P_{center_2} is the central pressure for the Gaussian spatial distribution, R is 0.8 which is the rising rate of the Gaussian spatial distribution, and R_s is the radius of laser spot 1.3 mm. It can be calculated that:

$$V_2 = 2.95 P_{center\_2} \tag{2.21}$$

For lasers with a uniform spatial distribution, the central pressure of the shock wave is the peak pressure:

$$P_{peak\_gaussian} = P_{center\_1} \tag{2.22}$$

$$V_2 = 2.95 P_{peak\_gaussian} \tag{2.23}$$

Assuming that in LSP processing, the laser power densities of the laser with a uniform spatial distribution and a Gaussian spatial distribution are the same, the input energy shall also be the same under the same spot diameter, so the volume of spatial distribution can be deemed as the same, and so:

$$V_1 = V_2 \tag{2.24}$$

$$P_{peak\_gaussian} = 1.8 P_{peak\_ceiling} \tag{2.25}$$

$$P_{peak\_ceiling} = 1.94\sqrt{I_0} \tag{2.26}$$

References

1. S. Chupakhin, B. Klusemann, N. Huber, N. Kashaev, Application of design of experiments for laser shock peening process optimization. Int. J. Adv. Manuf. Technol. (2019)
2. M. Tsuyama, N. Ehara, K. Yamashita, M. Heya, H. Nakano, Effect of laser peening with glycerol as plasma confinement layer. Appl. Phys. A **124**(3) (2018)
3. S. Zabeen, M. Preuss, P.J. Withers, Evolution of a laser shock peened residual stress field locally with foreign object damage and subsequent fatigue crack growth. Acta Mater. **83**, 216–226 (2015)
4. R. Fabbro, J. Fournier, P. Ballard, D. Devaux, J. Virmont, Physical study of laser-produced plasma in confined geometry. J. Appl. Phys. **68**(2), 775–784 (1990)
5. E. Cuenca, M. Ducousso, A. Rondepierre, L. Videau, N. Cuvillier, L. Berthe, F. Coulouvrat, Propagation of laser-generated shock waves in metals: 3D axisymmetric simulations compared to experiments. J. Appl. Phys. **128**(24) (2020)
6. P. Peyre, C. Carboni, A. Sollier, L. Berthe, C. Richard, E.D.L. Rios, R. Fabbro, New trends in laser shock wave physics and applications. High-power Laser Ablation IV (2002)
7. M. Malinauskas, A. Zukauskas, S. Hasegawa, Y. Hayasaki, V. Mizeikis, R. Buividas, S. Juodkazis, Ultrafast laser processing of materials: from science to industry. Light Sci. Appl. **5** (2016)
8. A.F. Kockum, A. Miranowicz, S. De Liberato, S. Savasta, F. Nori, Ultrastrong coupling between light and matter. Nat. Rev. Phys. **1**(1), 19–40 (2019)
9. S. Mondal, V. Narayanan, W.J. Ding, A.D. Lad, B.A. Hao, S. Ahmad, W.M. Wang, Z.M. Sheng, S. Sengupta, P. Kaw, A. Das, G.R. Kumar, Direct observation of turbulent magnetic fields in hot, dense laser produced plasmas. Proc. Natl. Acad. Sci. U.S.A. **109**(21), 8011–8015 (2012)
10. D.S. Dovzhenko, S.V. Ryabchuk, Y.P. Rakovich, I.R. Nabiev, Light-matter interaction in the strong coupling regime: configurations, conditions, and applications. Nanoscale **10**(8), 3589–3605 (2018)
11. S. Jiang, L.L. Ji, H. Audesirk, K.M. George, J. Snyder, A. Krygier, P. Poole, C. Willis, R. Daskalova, E. Chowdhury, N.S. Lewis, D.W. Schumacher, A. Pukhov, R.R. Freeman, K.U. Akli, Microengineering laser plasma interactions at relativistic intensities. Phys. Rev. Lett. **116**(8) (2016)
12. R. Ramis, K. Eidmann, J. Meyer-ter-Vehn, S. Huller, Multi-fs—A computer code for laser-plasma interaction in the femtosecond regime. Comput. Phys. Commun. **183**(3), 637–655 (2012)
13. L. Berthe, R. Fabbro, P. Peyre, L. Tollier, E. Bartnicki, Shock waves from a water-confined laser-generated plasma. J. Appl. Phys. **82**(6), 2826–2832 (1997)
14. A. De Giacomo, M. Dell'Aglio, O. De Pascale, R. Gaudiuso, V. Palleschi, C. Parigger, A. Woods, Plasma processes and emission spectra in laser induced plasmas: a point of view. Spectrochim. Acta Part B-At. Spectro. **100**, 180–188 (2014)
15. S. Zhu, Y.F. Lu, M.H. Hong, X.Y. Chen, Laser ablation of solid substrates in water and ambient air. J. Appl. Phys. **89**(4), 2400–2403 (2001)
16. A.K. Gujba, M. Medraj, Laser peening process and its impact on materials properties in comparison with shot peening and ultrasonic impact peening. Materials **7**(12), 7925–7974 (2014)
17. V.H. Whitley, S.D. McGrane, D.E. Eakins, C.A. Bolme, D.S. Moore, J.F. Bingert, The elastic-plastic response of aluminum films to ultrafast laser-generated shocks. J. Appl. Phys. **109**(1) (2011)
18. M.A. Barrios, D.G. Hicks, T.R. Boehly, D.E. Fratanduono, J.H. Eggert, P.M. Celliers, G.W. Collins, D.D. Meyerhofer, High-precision measurements of the equation of state of hydrocarbons at 1–10 Mbar using laser-driven shock waves. Phys. Plasmas **17**(5) (2010)
19. F. Fiuza, A. Stockem, E. Boella, R.A. Fonseca, L.O. Silva, D. Haberberger, S. Tochitsky, W.B. Mori, C. Joshi, Ion acceleration from laser-driven electrostatic shocks. Phys. Plasmas **20**(5) (2013)

20. J.P. Knauer, O.V. Gotchev, P.Y. Chang, D.D. Meyerhofer, O. Polomarov, R. Betti, J.A. Frenje, C.K. Li, M.J.E. Manuel, R.D. Petrasso, J.R. Rygg, F.H. Seguin, Compressing magnetic fields with high-energy lasers. Phys. Plasmas **17**(5) (2010)
21. S.S. Harilal, B.E. Brumfield, B.D. Cannon, M.C. Phillips, Shock wave mediated plume chemistry for molecular formation in laser ablation plasmas. Anal. Chem. **88**(4), 2296–2302 (2016)
22. K.H. Kurniawan, M.O. Tjia, K. Kagawa, Review of laser-induced plasma, its mechanism, and application to quantitative analysis of hydrogen and deuterium. Appl. Spectrosc. Rev. **49**(5), 323–434 (2014)
23. B. Arman, S.N. Luo, T.C. Germann, T. Cagin, Dynamic response of Cu46Zr54 metallic glass to high-strain-rate shock loading: Plasticity, spall, and atomic-level structures. Phys. Rev. B **81**(14) (2010)
24. X.H. Zou, B. Lu, W. Pan, L.S. Yan, A. Stohr, J.P. Yao, Photonics for microwave measurements. Laser Photonics Rev. **10**(5), 711–734 (2016)
25. W. Garen, F. Friebel, V. Braun, S. Koch, U. Teubner, Laser-induced shock waves from micro-scale volumina and in small tubes. Shock Waves **22**(4), 281–286 (2012)
26. Z. Henis, S. Eliezer, E. Raicher, Collisional shock waves induced by laser radiation pressure. Laser Part. Beams **37**(3), 268–275 (2019)
27. S. Eliezer, J.M. Martinez-Val, Z. Henis, N. Nissim, S.V. Pinhasi, A. Ravid, M. Werdiger, E. Raicher, Physics and applications with laser-induced relativistic shock waves. High Power Laser Sci. Eng. **4** (2016)
28. J.X. Wang, X. Gao, C. Song, J.Q. Lin, Experimental study of shock waves induced by a nanosecond pulsed laser in copper target. Acta Physica Sinica **64**(4) (2015)
29. R. Zhao, R.Q. Xu, Z.C. Liang, Laser-induced plasma shock wave propagation underwater. Optik **124**(12), 1122–1124 (2013)
30. I.I. Oleynik, B.J. Demaske, V.V. Zhakhovsky, N.A. Inogamov, C.T. White, MD simulations of laser-induced ultrashort shock waves in nickel (2011)
31. K. Yuan, Y. Sumi, Simulation of residual stress and fatigue strength of welded joints under the effects of ultrasonic impact treatment (UIT). Int. J. Fatigue **92**, 321–332 (2016)

Chapter 3
Gradient Microstructure Characteristics and the Formation Mechanism in Titanium Alloy Subjected to LSP

3.1 Introduction

Titanium alloy has a series of advantages, such as high strength, good corrosion resistance, good medium temperature performance, etc. It is often used in the manufacture of compressor blades/disks for aero-engines and other components to reduce weight and improve the thrust-weight ratio. However, due to the extreme operation environment of titanium alloy fans and compressor blades, especially the extremely high rotating speed of the rotor blades (over 12,000 rpm), the tangential velocity at the blade tips is as high as 450–500 m/s, causing great centrifugal inertia force, accompanied by vibration stress caused by airflow micro-vibration, which easily leads to fatigue. In addition, during take-off and landing, it is very easy for the engine to inhale hard, foreign objects such as gravel and metal fragments, which will damage the front blades of the fan and compressor, form fatigue-weakened areas and eventually lead to fatigue fracture, which is one of the main manifestations of aero-engine "heart disease".

Since the maximum working stress and fatigue crack initiation are generally located on the surface of components, the LSP method is used to improve the surface state of fan and compressor blades to reduce the risk of fatigue failure caused by service damage. In 2004, in the "Engine Structural Integrity Program (MIL-HD BK-1783 BW/CHANG2)", the United States put forward the requirement of changing the surface state of fan and compressor blades by LSP and shot peening, and explicitly incorporated it into the engine structural design, which greatly improves the fatigue performance of F101 engine compressor's titanium alloy blades after being damaged by hard objects, increases the foreign object damage (FOD)tolerance by more than 15 times, and greatly reduces the maintenance cost of the engine. At present, this surface treatment technology has been applied to titanium alloy blades of fan/compressor of engines such as F101, F110, F404, F119, F135, etc., which effectively solves the problem of insufficient resistance to hard objects. Many scholars have carried out relevant research, for example, Spanrad et al. [1] studied the variation law of

© Zhejiang University Press 2021
L. Zhou and W. He, *Gradient Microstructure in Laser Shock Peened Materials*, Springer Series in Materials Science 314,
https://doi.org/10.1007/978-981-16-1747-8_3

fatigue strength of FOD samples after LSP treatment, and the results show that fatigue cracks of strengthened samples do not easily occur and the propagation rate is obviously reduced. Lin et al. [2] has carried out tests of LSPed samples impacted by foreign objects at 0° and 45° angles, and studied the influence of different angles of impact on fatigue performance of FOD samples before and after LSP. The results show that a 45° impact has greater influence on fatigue performance. Zabeen et al. [3] analyzed the distribution of the field of stress on the damaged parts of LSPed samples, and measured the residual stress on the damaged parts by the synchrotron radiation method. The results show that the stress generated by FOD and LSP will be coupled to a certain extent, the residual tensile stress at the edge of the damaged notch will be greatly reduced, and the residual compressive stress at the bottom of the damaged notch will be increased. Both of them can inhibit the propagation of fatigue cracks and improve fatigue performance. They also tested the residual stress by the X-ray diffraction method. The results show that the residual compressive stress introduced by LSP can effectively inhibit crack propagation and improve the fatigue strength of materials. Zhang et al. [4] studied the influence of different impact times on fatigue properties of TC4 titanium alloy. The increase in residual compressive stress caused by the increase in impact times greatly improves fatigue strength, but it is not unlimited, and fatigue performance is also related to other factors. Nie et al. [5] studied the vibration fatigue of TC6 titanium alloy subjected to LSP, and found that LSP increases the fatigue limit of TC6 standard vibration specimen from 438.6 to 526.7 MPa, which is attributed to the refinement of the microstructure and to the high residual compressive stress caused by LSP. It has been found that the most current research is mainly on the improvement of fatigue performance by LSP forming residual compressive stress on the surface of the titanium alloy blade, and less on the change in the gradient microstructure and the influence mechanism of fatigue performance. Therefore, in the third and fourth chapters of this book, the characterization of the gradient microstructure, the mechanism of formation and the mechanism of improving the fatigue performance of titanium alloy by LSP are introduced emphatically.

3.2 Residual Stress Characteristics of LSP in Titanium Alloy

3.2.1 Experiments and Methods

TC4 titanium alloy selected in this section is an $\alpha + \beta$ type dual-phase alloy with medium strength, and its chemical element composition is shown in Table 3.1, and which has the characteristics of high reliability and good structural benefit. The basic physical parameters are shown in Table 3.2; the heat treatment mechanism of TC4 is according to the technical standard GJB 494-1988, and the basic performance is shown in Table 3.3.

Table 3.1 Chemical composition of TC4 titanium alloy (%)

Alloying element			Impurity not more than						
Al	V	Ti	Fe	C	N	H	O	Other elements	
								Single	Sum
5.5–6.8	3.5–4.5	Bal.	1.6–2.4	1.6–2.4	0.05	0.0125	0.13	0.1	0.4

Table 3.2 Physical parameters of TC4 titanium alloy

Material	Density/g cm^{-3}	Poisson's ratio	Elastic model/GPa	Shear modulus/GPa	Bulk modulus/GPa
TC4	4.44	0.34	109	44	119

Table 3.3 Heat treatment mechanism and basic mechanical properties of TC4 titanium alloy

Technical standard	Heat treatment	σ_b/MPa	$\sigma_{0.2}$/MPa	δ_5/%
GJB 494-1988	Annealing: 700–850 °C, 0.5–2 h, air cooling	925	870	12

The residual stress field of TC4 titanium alloy treated by LSP was measured by the LXRD residual stress tester manufactured by PROTO, as shown in Fig. 3.1. The scanning method is the same-incline fixed ψ, θ-θ symmetric scanning method, the target material is Cu target, the radiation type is Cu-Kα, the diffraction angle 2θ is 139°–142°, the direction of diffraction crystal plane is {213}, and the diffraction

Fig. 3.1 LXRD X-ray diffractometer

wavelength is 1.541 nm; when testing residual stress at different depths, an 8818 V^{-3} electrolytic polisher is used to remove surface materials layer by layer.

3.2.2 The Law of the Influence of Laser Power Density

The residual stress distribution on the surface of TC4 titanium alloy treated by LSP is shown in Fig. 3.2. In this book, the residual stress field distribution of the three power densities 3.94, 5.26 and 6.57 GW/cm^2 are measured under flat-topped and Gaussian spatial energy distribution. It can be seen from Fig. 3.3 that when the spatial energy type is Gaussian distribution, the average residual stress on the surface is −656, −778 and −826 MPa. With the increase in power density, the residual compressive stress on the surface of titanium alloy strengthened by LSP increases gradually, and the rate of increase decreases with the increase in power density.

When the type of spatial energy is flat-topped distribution, the surface residual stresses are −565 MPa, −658 MPa and −733 MPa respectively. The residual compressive stress on the surface also increases with the increase in power density, and the amplitude becomes smaller and smaller. This is because with the increase in power density, the energy received by the material surface increases, and the plastic deformation of the surface increases, so the residual compressive stress correspondingly increases.

The processing parameters under the two spatial energy distributions are identical, and the residual compressive stress produced by the laser of Gaussian spatial distribution is larger than that of flat-topped distribution. This is because the peak pressure of the laser beam with Gaussian spatial distribution is higher, while the spatial energy of that with flat-topped distribution is more uniform. When the power density is equal, it is assumed that the absorption of light energy is completely the same and

Fig. 3.2 Influence of power density on the surface residual stress field distribution

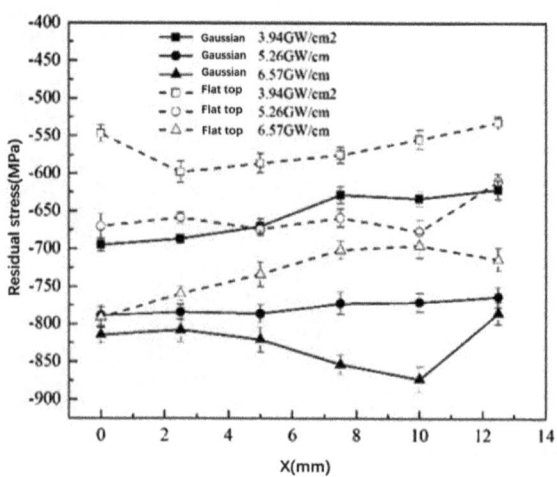

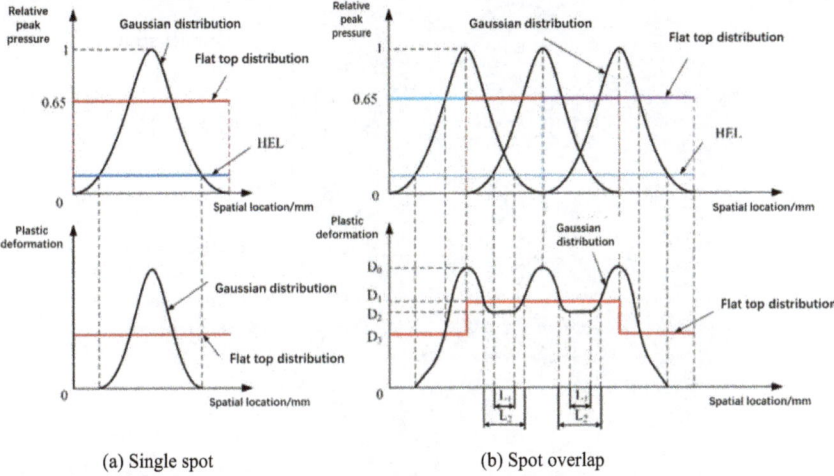

(a) Single spot   (b) Spot overlap

Fig. 3.3 Relationship between shock wave pressure and the plastic deformation of different spatial energy types

the energy is conserved. It can be obtained that the peak pressure of the laser beam in Gaussian spatial distribution is 1.5 times that in the flat-topped distribution [6]. The relationship between shock wave pressure and plastic deformation of a single spot and of an overlapping spot is shown in Fig. 3.3. It can be seen from Fig. 3.3 that the plastic deformation of the laser beam of Gaussian spatial distribution is larger than that of flat-topped distribution under the condition of spot overlap. The process of the generation of plastic deformation of the Gaussian-spatial-distribution laser beam is represented by black lines in Fig. 3.3. At the edge of the circular light spot, the shock wave pressure is less than the Hugoniot elastic limit of TC4 titanium alloy σ_{HEL}, where the material only undergoes elastic deformation. Hugoniot elastic limit refers to σ_{HEL} the minimum stress that causes dynamic response and dynamic plastic deformation of materials. Near the center of the light spot, the pressure produced by the shock wave of the Gaussian spatial distribution laser is greater than σ_{HEL}, where the material undergoes severe plastic deformation, and the deformation increases with the increase in the shock wave pressure. At the center of the first spot, the shock wave pressure induced by laser reaches the maximum value, while the shock wave pressure of the second spot increases from the edge (i.e., the center of the first spot) due to the overlap ratio of 50%. When the pressure generated by the shock wave in the second light spot exceeds σ_{HEL}, there are two pressures in the overlapping area that cause the deformation of TC4 titanium alloy to act on the material surface successively, and the deformation of the material after superposition is as shown in Fig. 3.3b L1 section.

When the spatial energy of the laser is flat-topped, the shock wave pressure is always greater than the dynamic elastic limit of the material, so plastic deformation will occur at any position in the strengthening area, and the change in deformation

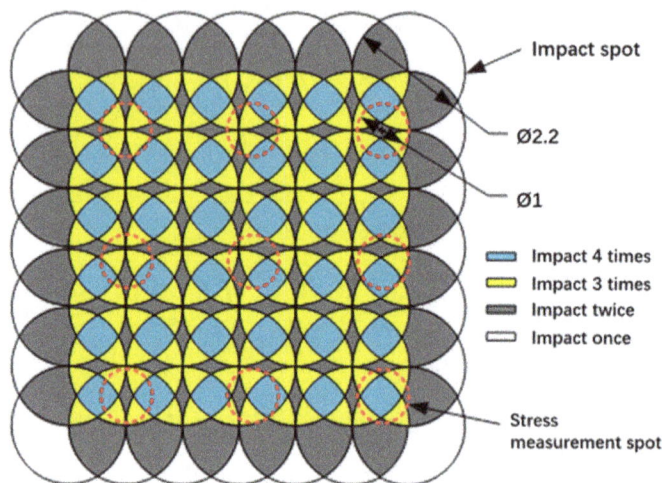

Fig. 3.4 Average impact times and measurement circle distribution in different spot areas

with the spot position is shown by the red line in Fig. 3.3. Essentially, the overlapping area of the light spot gets more impacts on average, and its schematic diagram is shown in Fig. 3.4. The difference in average impact times caused by the overlapping rate in different positions in a single circular light spot makes the material deformation not proportional to the impact times. The research shows that the increase in plastic deformation on the material surface decreases with the increase in impact times [7]. Therefore, the deformation of the overlapping area is less than twice that of the non-overlapping area, that is, $D_1 < 2D_3$.

In the overlapping area, the deformation by the Gaussian distribution laser is smaller than that of the flat-topped distribution laser in the L_2 segment, but $D_0D_2 \gg D_1D_2$, and the plastic deformation at other positions is larger than that by the flat-topped distribution laser, so the average plastic deformation by the Gaussian distribution laser is larger than that by the flat-topped distribution laser. Therefore, when the LSP process and parameters are the same, the average residual compressive stress on the surface with Gaussian spatial energy distribution is larger than that with the flat-topped distribution.

Therefore, when the power density is the same, the residual compressive stress induced by the laser beam with Gaussian spatial energy distribution is larger than that by the laser beam with the flat-topped distribution.

The residual stress distribution in depth is shown in Fig. 3.5, which is measured by gradual delamination, and the average value measured three times in the same horizontal line is taken as the final value. When the spatial energy distribution is a Gaussian distribution, the residual compressive stress affected layer of 0.565 mm is produced when the power density is 3.94 GW/cm^2. When the power density is 5.26 GW/cm^2, a residual compressive stress affected layer of 1.063 mm is produced. When the power density is 6.57 GW/cm^2, the residual compressive stress affected

Fig. 3.5 Influence of power density on residual stress distribution in depth

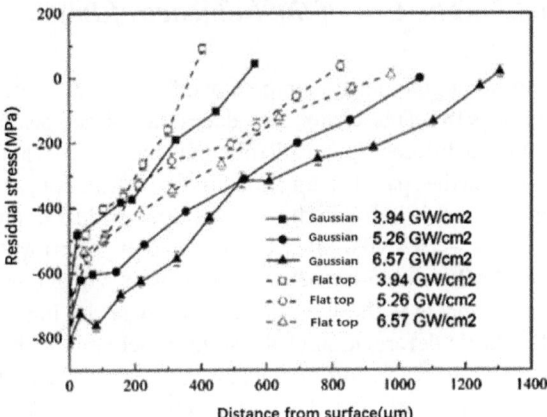

layer reaches 1.305 mm. It can be seen from Fig. 3.5 that with the increase in power density, the residual compressive stress affected layer in depth gradually becomes larger, and the increasing range becomes smaller and smaller. At the same depth, the higher the power density, the greater the residual compressive stress. This is because higher power density produces higher peak pressure in the LSP process. With the increase in depth, the shock wave gradually attenuates in the material, while the higher the peak pressure, the slower the shock wave attenuates in titanium alloy, and the greater the peak pressure and residual compressive stress at the same depth.

When the spatial energy distribution is flat-topped, the residual compressive stress affected layers of 0.405 mm, 0.825 mm, and 0.976 mm in depth are respectively produced by LSP with three power densities. With the increase in power density, the depth of the residual compressive stress layer increases, but the range of increase decreases. At the same depth, the residual compressive stress increases slightly with an increase in power density. At the same power density, the residual compressive stress affected layer produced by a flat-topped distribution laser is smaller than that produced by a Gaussian distribution laser, because the peak pressure of the flat-topped distribution laser is smaller than that of the Gaussian distribution laser at the same power density, so the propagation depth of the shock wave inside the material is smaller than that of the Gaussian distribution laser, and its residual compressive stress layer is smaller than that of the Gaussian distribution laser.

To sum up, from the average value of the residual compressive stress field, no matter how the power density changes, the residual stress field produced by the laser beam with Gaussian spatial energy distribution is better than that produced by the flat-topped distribution.

3.2.3   The Law of the Influence of Impact Times

Impacting times have an important effect on the residual stress field on titanium alloy surfaces [8]. This section introduces the surface residual stress distribution by LSP with two different spatial distribution types for 1, 3, and 5 times, as shown in Fig. 3.6.

When the spatial energy is Gaussian distribution, and the LSP parameters are kept unchanged, it is found that the average residual compressive stress on the surface is about −757 MPa after 1 impact, about −850 MPa after 3 impacts, and about −800 MPa after 5 impacts. The value of residual compressive stress on the surface by 3 impacts is the largest, because when the impacts increase from 1 to 3, and the plastic deformation of the surface increases with the increase in the number of impacts, so the residual compressive stress on the surface also increases. When the impacts increase from 3 to 5, due to the higher peak pressure of the laser beam with Gaussian spatial distribution, the surface plastic deformation has been saturated due to multiple impacts, and the surface plastic waves no longer continue to propagate downward, but produce reverse rarefaction waves acting on the material surface, which makes the surface residual compressive stress value decrease after 5 impacts.

When the distribution of spatial energy is flat-topped, the average residual compressive stress on the surface produced by the 1, 3 and 5 impacts is about −670 MPa, −784 MPa and −792 MPa, respectively, showing an increasing trend with the increase in the number of impact times. It can be seen from Fig. 3.6 that the average residual compressive stress produced by 3 impacts increases greatly compared with that by 1 impact, while the rate of increase rate of 5 impacts is smaller that of 3 impacts. This is also caused by the plastic saturation phenomenon accompanying the increase in the number of impact times. However, for the laser with flat-topped distribution, the surface residual stress does not decrease after 5 impacts, which is because the peak pressure of the laser beam with flat-topped distribution is smaller than that of Gaussian distribution.

Fig. 3.6 Influence of impact times on the surface residual stress field distribution

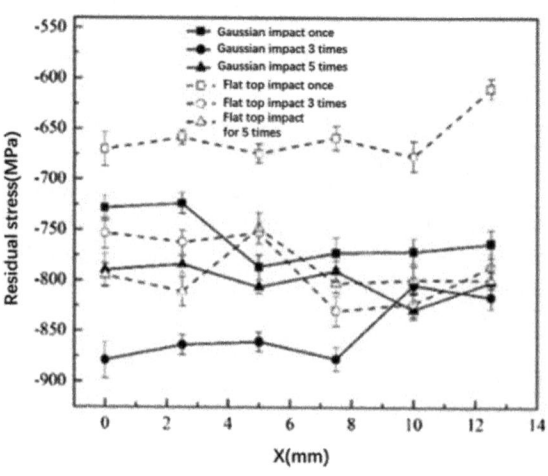

Fig. 3.7 Influence of impact
times on residual stress
distribution in depth

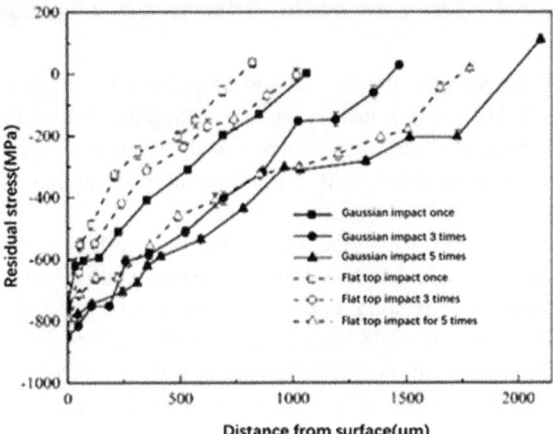

The distribution of residual stress in depth is shown in Fig. 3.7 when the spatial energy is Gaussian, the residual compressive stress affected layer of 1.063 mm is produced after 1 impact, 1.47 mm after 3 impacts, and 2.1 mm after 5 impacts, increasing with the increase in the number of impact times. Unlike the surfaceresidual stress distribution, the deepest residual compressive stress affected layer occurs after 5 impacts. In the area below the plastic deformation saturation, multiple impacts make the shock wave propagate deeper, and the depth of the residual compressive stress affected layer increases with the increase in the number of impact times.

For the laser beam of flat-topped spatial distribution, the depth of the residual compressive stress affected layer follows the same rule as that of Gaussian distribution. Residual compressive stress affected layers of 0.825, 1.019, and 1.783 mm are produced under different impact times, which indicates that the depth increases with the increase in the number of impact times. When the impact times of each spatial distribution laser are the same, the depth of the residual compressive stress affected layer of the Gaussian distribution laser is larger than that of the flat-topped distribution laser, because the peak pressure of the shock wave of the Gaussian spatial distribution laser is larger than that of the flat-topped distribution laser, and its shock wave propagates deeper, causing plastic deformation of the deeper material, resulting in a deeper residual compressive stress affected layer. Some researchers [9–11] also found that the greater the peak pressure a shock wave has, the longer the time it takes for the pressure to decay below the yield strength of the material, the deeper the shock wave propagates, and the deeper the residual compressive stress affected layer is.

To sum up, appropriate impact times can avoid the problem of residual stress reduction due to the reverse action of surface rarefaction waves, but the residual compressive stress affected layer gradually deepens with the increase in the number of impact times. Therefore, the laser beam with Gaussian spatial distribution can produce a deeper field of residual compressive stress than that with flat-topped distribution under suitable impact times.

3.2.4 The Law of the Influence of Overlapping Rate

The overlapping rate also has a significant influence on the surface residual stress field [12]. In this section, three overlapping rates of 15%, 30%, and 50% are adopted respectively, and the influence of different overlapping rates on the distribution of the residual stress field of titanium alloy strengthened by LSP with different spatial energy distributions is introduced.

Figure 3.8 shows the influence of the overlap ratio on the distribution of the field of surface residual stress. When the spatial energy distribution is Gaussian distribution, the average residual compressive stress produced by LSP with the overlap ratio of 15% is about −548 MPa, for the overlap ratio of 30%, it is about −625 MPa, and for the overlap ratio of 50%, it is about −757 MPa. The surface residual stress increases with the increase in the overlap ratio. This is because the energy distribution in the laser beam spot with Gaussian spatial distribution is uneven, and when the surface is peened by a single spot, only the central position in the spot will produce plastic deformation, and the edge of the spot will not produce plastic deformation, as shown in Fig. 3.3a, and tensile stress will easily occur after LSP. Therefore, the overlapping of laser spots will make the positions with a lower shock wave pressure at the edge of light spots overlap many times, and the plastic deformation and residual compressive stress at the edge will increase. In addition, some literature [13] believes that the essence of spot overlapping is the increase in the number of impact times in the overlapping area, and the higher the overlapping rate, the more impact times in a single spot, so the greater the value of residual compressive stress.

When the spatial energy distribution is flat-topped, the average values of residual compressive stress under three overlapping rates are −622 MPa, −608 MPa and −658 MPa, respectively, with little difference. This is because the distribution of spatial energy in a flat-topped beam spot is relatively uniform, and the residual compressive stress in a single spot is also relatively uniform, and plastic deformation

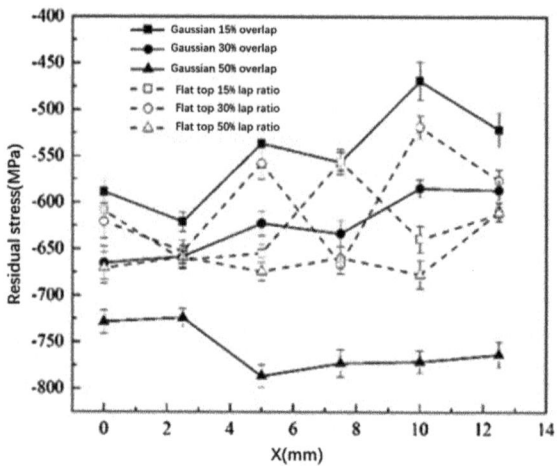

Fig. 3.8 Effect of lap ratio on the surface residual stress distribution

will occur in the whole spot. At the same time, the greater the number of impact times, the greater the possibility of plastic saturation. Therefore, the increase in residual compressive stress on the surface is not obvious after increasing the overlapping rate. Literature [14] also shows that the overlapping rate of 10% is the best in terms of economy and impact effect, and it is concluded that the enhancement effect of the flat-topped laser beam is not obvious with an increase in overlapping rate.

When the overlap rate is 15%, the surface residual stress field produced by a flat-topped distribution laser is better than that produced by a spatial Gaussian distribution laser. At a 30% overlap rate, the field of surface residual stress produced by a spatial Gaussian distribution laser and a flat-topped distribution laser is basically the same, but the impact of the laser of the Gaussian distribution laser at the overlap rate of 50% will produce the maximum residual compressive stress.

The residual stress field distribution in depth is shown in Fig. 3.9. When the spatial energy distribution is Gaussian distribution, laser beams at the overlapping rate of 15 and 30% produce the same residual compressive stress depth, which is about 0.85 mm, and the residual compressive stress affected layer reaches 1 mm under the overlapping rate of 50%. When the spatial energy distribution is flat-topped, the depth of the residual compressive stress layer under three overlapping rates is 0.564 mm, 0.663 mm, and 0.825 mm respectively. With the increase in the overlapping rate, the depth of residual compressive stress increases. Under each of the three overlapping rates, the residual compressive stress layer produced by the Gaussian spatial distribution laser is larger than that of the flat-topped distribution laser, because the peak pressure of the Gaussian spatial distribution laser is larger than that of the flat-topped distribution laser, and the higher the peak pressure, the deeper the shock wave propagates in the material and the deeper the residual compressive stress layer.

To sum up, the overlapping rate has a significant influence on the residual stress field under two different energy spatial distributions. The residual compressive stress field on the surface and in depth produced by the Gaussian spatial distribution laser

Fig. 3.9 Effect of lap ratio on residual stress distribution of the direction of the depth

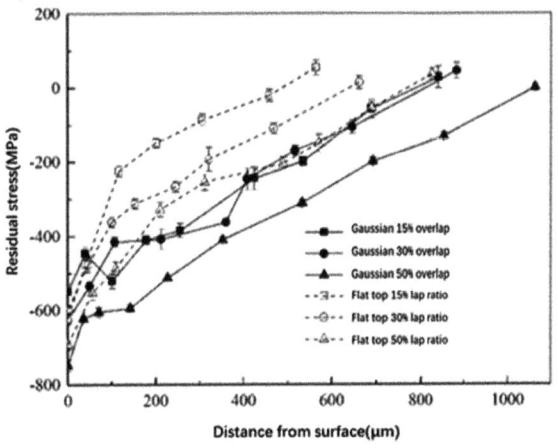

can reach the maximum at 50% overlapping rate; the residual compressive stress on the surface caused by the flat-topped distributed laser beam has little change with the different overlapping rates, but the depth of the residual stress field changes noticeably with the increase in overlapping rate. The optimal residual compressive stress field can be obtained by using Gaussian spatial energy distribution with a 50% overlapping rate.

3.2.5 Stability of Residual Stress Induced by LSP Under Thermal and Mechanical Load

(1)   Study on the law of residual stress relaxation under mechanical loading

The fatigue limit of materials can be improved by introducing a certain amount of residual compressive stress by LSP, but the residual compressive stress is unstable during the fatigue process, and under a certain external load, the residual stress will have a relaxation effect [15]. In this section, the law of the relaxation of the residual stress field under different load peaks under certain stress ratios is introduced. The fatigue test is carried out under the stress ratio of 0.1 and the cycle frequency of 1 Hz. The maximum static load and vibration load under 500 MPa are 14.33 kN and 7.88 kN respectively. After a certain number of tensile and compressive tests, the residual stress on the surface and in depth is measured by an X-ray diffractometer. The relaxation curve of surface residual stress under a 500 MPa alternating cycle load is shown in Fig. 3.10.

It can be seen from Fig. 3.10 that during the fatigue cycle, the residual stress produced by LSP is obviously relaxed. The value of initial surface residual compressive stress is −665 MPa, which decreases by about 100 to −567 MPa after one cycle, and decreases to −530 MPa after five cycles. When the number of cycles reaches 10, the residual stress is basically stable at about −500 MPa, and the degree

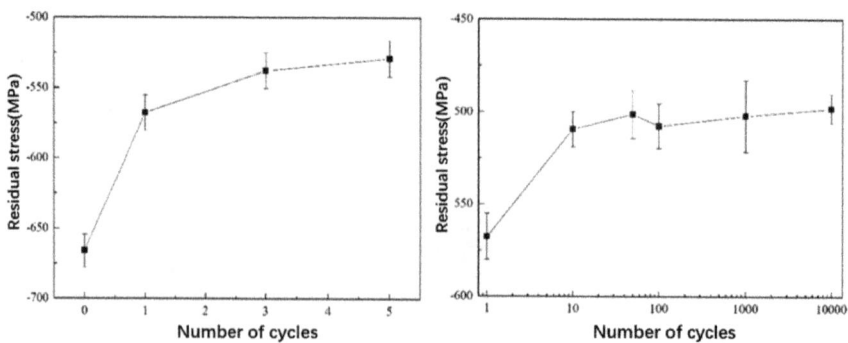

Fig. 3.10 Surface residual stress relaxation curves under fatigue load of 500 MPa and different cycles

of relaxation of residual stress reaches 24.8%. The relaxation of residual stress under alternating load mainly occurs in the first few fatigue cycles of alternating load, and the relaxation of stress of the first five cycles is the most serious.

During the 10th–10,000th cycles, the degree of relaxation of residual stress is relatively small, and it is basically in a stable state. Stable residual stress is considered as the key factor to improve fatigue life. The effective residual stress σ_{rs}^{eff}:

$$\sigma_{rs}^{eff} = \sigma_{rs} - \sigma_{rs}^{relax} \tag{3.1}$$

The relaxation of residual compressive stress is the result of the superposition of alternating external load and residual stress after LSP inside the material. When plastic deformation occurs (such as the loading process in a fatigue test), the internal stress of the material will be redistributed, which will lead to the relaxation of residual compressive stress.

$$\sigma_{appl} = \sigma_f^{ten} - \sigma_{rs}^{bal} \tag{3.2}$$

In the above Formula (3.2), σ_{appl} is the actual stress in the process of the alternating cyclic load, $\sigma_f^{tensile}$ is tensile stress, and σ_{rs}^{bal} is equilibrium stress. The relaxation of the residual stress of titanium alloy treated by LSP basically occurs in the first few cycles, because when the actual stress in the material is greater than its yield strength, plastic deformation will occur, and plastic deformation determines the distribution of stress in the material, so the stress relaxation phenomenon appears. In the later alternating load cycle, because the stress relaxation in the early stage has redistributed the residual stress in the material, the actual stress at this time cannot lead to plastic deformation, and all of them are mainly elastic deformations, so the residual stress will not relax in the later cycle loading.

Figure 3.11 shows the law of residual stress relaxation under a fatigue load of 400 and 300 MPa. It can be seen from the figure that under the action of a 400 MPa alternating load, the residual compressive stress on the surface of TC4 titanium alloy

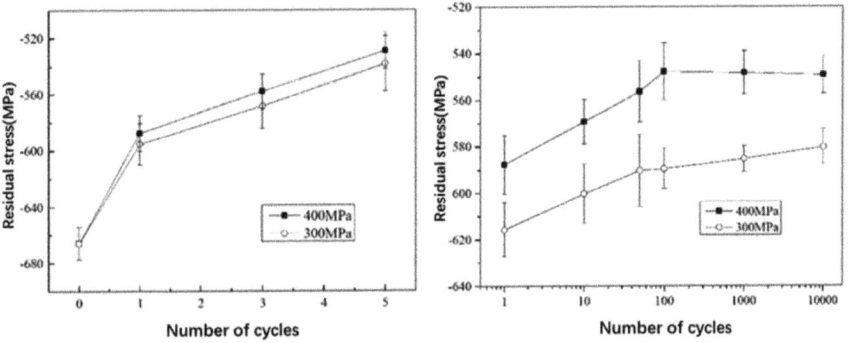

Fig. 3.11 Relaxation curves for surface residual stress under different fatigue loads

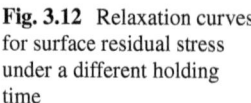

Fig. 3.12 Relaxation curves for surface residual stress under a different holding time

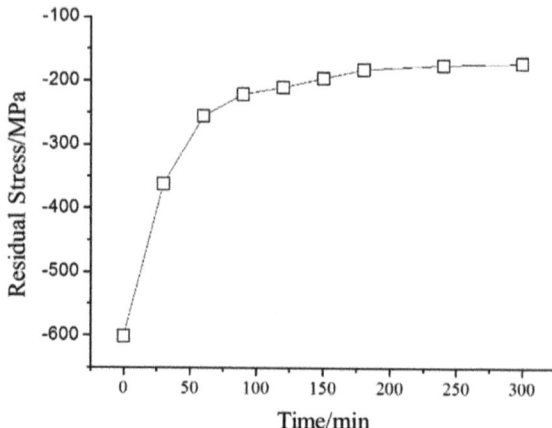

decreases from −665 to −587 MPa after one cycle. Under the action of a 300 MPa alternating load, the residual compressive stress decreases to −595 MPa after one cycle. When the surface stress is the same, the greater the alternating load is, the more obvious the relaxation of residual stress is. After 1,000 cycles under the two kinds of loads, the stress also tends to be basically stable.

(2)   The law of the relaxation of residual stress in a thermal environment

The residual stress will also have stress relaxation under the action of high thermal stress, but the mechanism is different from mechanical relaxation. Thermal stress relaxation is caused by the activation and movement of crystal defects such as stacking fault and twins in materials under the action of thermal stress, which makes the lattice distortion of materials recover. Figure 3.12 is the distribution curve of residual stress of the TC4 titanium alloy sheet after LSP at 400 °C for 5 h. LSP parameters: power density 4.3 GW/cm^2, one impact time, overlapping rate 50%. The residual compressive stress on the surface is reduced from −601 to −255 MPa and the degree of stress relaxation is 57%. After 90 min, the stress distribution curve tends to be flat.

3.3   Characteristics of Gradient Microstructure of LSP in Titanium Alloy

3.3.1   Experiments and Methods

TC6 titanium alloy selected in this section is a martensite $\alpha + \beta$ two-phase titanium alloy with good comprehensive properties [16, 17], which is made up of a hexagonal close-packed (HCP) substrate α phase and body-centered-cubic (BCC) strengthened β phase. See Table 3.4 for the chemical composition and see Table 3.5 for the mechanical properties.

Table 3.4 Chemical composition of TC6 titanium alloy (Wt%)

Alloying element	Al	Mo	Cr	Fe	Si	Ti
Component	5.5–7.0	2.0–3.0	0.8–2.3	0.2–0.7	0.15–0.40	Bal

Table 3.5 Mechanical properties of TC6 titanium alloy (longitudinal)

Brand number	Technology standard	HEL (MPa)	Room temperature instantaneous stretching				Room temperature shock		Hardness
			σ_b	$\sigma_{0.2}$	δ_5	ψ	α_{KV}	α_{KU}	HB
			(MPa)		(%)		(kJ/m²)		(kgf/mm²)
TC6	GB2965-87	2560	930	885	12	35	295	455	365

In this section, XRD, electron backscattering diffraction (EBSD), and TEM are used to systematically characterize the microstructure changes in titanium alloy under after LSP different parameters. Among them, TEM observation is the most important method of characterization to study the surface nanocrystallization of metal materials by LSP. TEM samples are prepared in two types: cross-sectional samples and thin-film samples at different depths. Cross-section samples are mainly used to determine the distribution of nano-microstructure inside the material and the process of the evolution of nanocrystals on the surface of metal materials after LSP. Thin-film samples are used to determine the microstructure distribution after observing cross-sectional samples.

3.3.2 XRD Phase

In order to obtain the characteristics of the microstructure of TC6 titanium alloy after LSP under different shock parameters, LSP tests with different power densities and different impact times are designed, and the gradient microstructure is analyzed from the aspects of phase composition, full width at half maximum of diffraction peak and so on. See Fig. 3.13 for the X-ray diffraction pattern of laser beams with the power density of 4.24 GW/cm² and 1, 3, 5, and 10 impacts respectively.

It can be seen from Fig. 3.13a that under different impact times, no new peaks appear in the XRD pattern of TC6 titanium alloy samples, and the phase composition is still α-Ti and β-Ti, indicating that no new phases are generated in the samples. After deducting the influence of instrument broadening, the Bragg diffraction peak is still broadened. According to the X-ray diffraction principle, the broadening of the Bragg diffraction peak is the joint effect of crystal grains refinement and micro stress (micro-strain), which shows that LSP increases the micro stress on the surface layer of the sample and refines the crystal grains. This is consistent with the results of general surface strengthening technology [18, 19]. The microstructure of titanium alloy formed by a single impact is characterized by a small size of an area of coherent

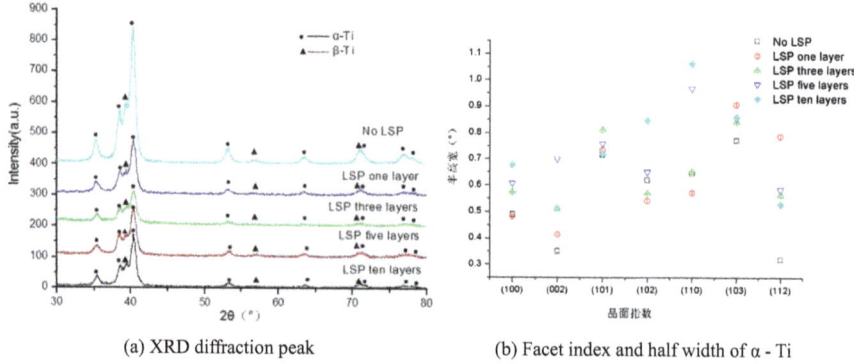

(a) XRD diffraction peak   (b) Facet index and half width of α - Ti

Fig. 3.13 XRD patterns of TC6 titanium alloy after different impact times

scattering and serious micro distortion of the crystal lattice. With the increase in the number of impact times, the level of lattice micro-distortion does not change greatly. It shows that the evolution of the microstructure in the deformation process is not related to the further refinement of the surface microstructure, but surface plastic deformation becoming more uniform and tending to saturation. Figure 3.13b shows the distribution of full width at half maxima (FWHM) of TC6 titanium alloy samples under different impact times. The FWHM of diffraction peaks of all crystal faces corresponding to α-Ti was broadened to varying degrees after LSP, and the FWHM of diffraction peaks on (100), (102) and (110) crystal faces increase with the increase in the number of impact times. The XRD patterns of TC6 titanium alloy with different power densities (2.83, 4.24 and 6 GW/cm^2) after 3 imapcts are shown in Fig. 3.14.

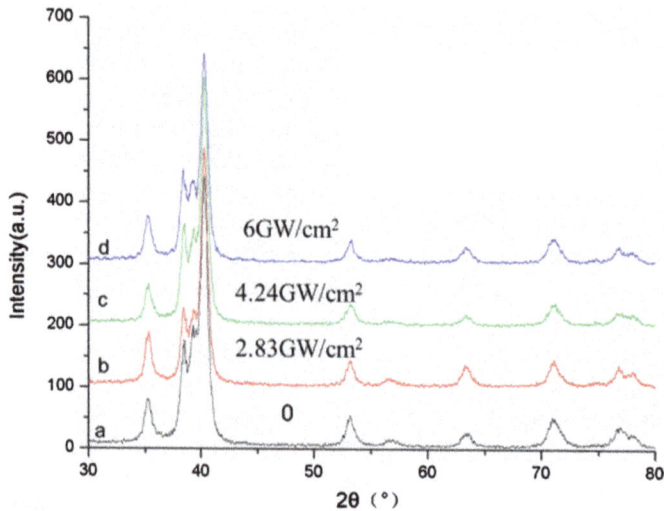

Fig. 3.14 XRD patterns of TC6 titanium alloy after LSP with different power densities

It can be observed from Fig. 3.14 that the Bragg diffraction peaks corresponding to each phase (α-Ti, β-Ti) are broadened after LSP. However, changing the power density does not cause the change in the position of the diffraction peak and the FWHM of TC6 titanium alloy. This shows that the change of power density does not change the phase distribution of TC6 titanium alloy, and has little effect on crystal grains refinement.

3.3.3 TEM Characterization

The microstructure of titanium alloy under different parameters of the LSP is characterized, and the influence of power density and impact times on the microstructure change is studied. TEM samples are processed by LSP with the parameters: one impact with 4.24 GW/cm^2, one impact with 6 GW/cm^2, and five impacts with 6 GW/cm^2. At first, observe the cross-sectional sample of titanium alloy after one impact with a power density of 4.24 GW/cm^2, as shown in Fig. 3.15.

It can be observed from Fig. 3.15a that LSP produces nanocrystals on the surface layer (area A) with a thickness of less than 1 μm. The electron diffraction diagram of the corresponding selected area (Fig. 3.15b) proves the existence of nanocrystals on the surface layer, and their orientations are randomly distributed. With the increase of depth (area B), the diffraction spot in the selected area (Fig. 3.15c) is obviously distorted. This shows that LSP produces high-density dislocations and large microscopic strain in the material.

Figure 3.16 is a high-resolution TEM image (HRTEM) of the surface layer area. It can be observed that a layer of amorphous structure with a thickness of about 10 nm is uniformly distributed above the nanocrystalline layer, and its atoms are disordered. A similar nano/amorphous mosaic structure was observed in some materials with grain refinement induced by strong plastic deformation. Related studies suggest that the mechanism of its formation is related to the plastic deformation at an ultra-high

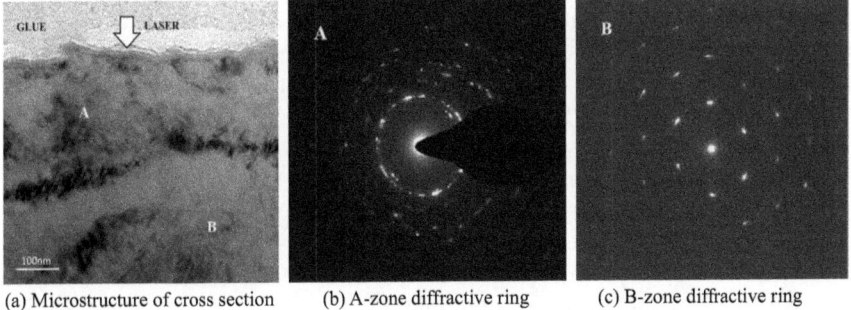

(a) Microstructure of cross section   (b) A-zone diffractive ring   (c) B-zone diffractive ring

Fig. 3.15 TEM images of the cross section of titanium alloy after one impact with power density of 4.24 GW/cm^2

Fig. 3.16 High resolution
HRTEM images of the
surface region

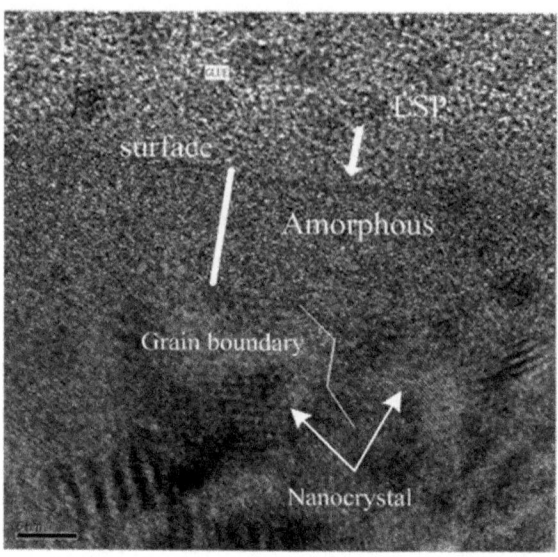

strain rate [20, 21]. The plastic deformation with an ultra-high strain rate induced by
LSP induces high dense dislocations and stacking faults in materials. High-density
dislocations distort the peripheral lattice, while the atomic arrangement is disordered
under the action of high-density stacking faults, resulting in an amorphous phase [20].
Some scholars believe that the amorphous phase is caused by the temperature rise
caused by LSP. That is to say, under the impact load, the local strong and severe
plastic deformation on the surface of the material causes the temperature to rise,
and then dynamical recrystallization under the rapid cooling of the substrate forms
amorphous structures or partial structures of nanocrystals. To study the influence of
laser power density, the laser power density is increased to 6 GW/cm^2, and the TEM
of the sample cross section after one shock is shown in Fig. 3.17.

It can be seen from Fig. 3.17a–b that local amorphous and nanocrystal structures
are formed on the surface of titanium alloy, and the thickness of the amorphous
layer is about 10 nm and that of nanocrystals is about 1 μm. The diffraction rings
corresponding to the three areas A, B, and C are shown in Fig. 3.17c–e, which are
characterized by an amorphous structure, a nanocrystal structure and a typical crystal
structure respectively. With the increase in laser power density, the thickness of the
nanocrystal layer has no obvious change.

In order to study the impact times, five LSP tests were carried out on TC6 alloy
under the same power density. Figure 3.18 shows the TEM image of the cross section
of the sample after 5 impacts.

After increasing the number of impact times, the thickness of the nanocrystal
layer is still 1 μm, and the diffraction rings corresponding to areas A and B show
large lattice distortion. At the same time, many grain boundaries that are not straight
lines can be seen from the microstructure images, which are curved and unequal in

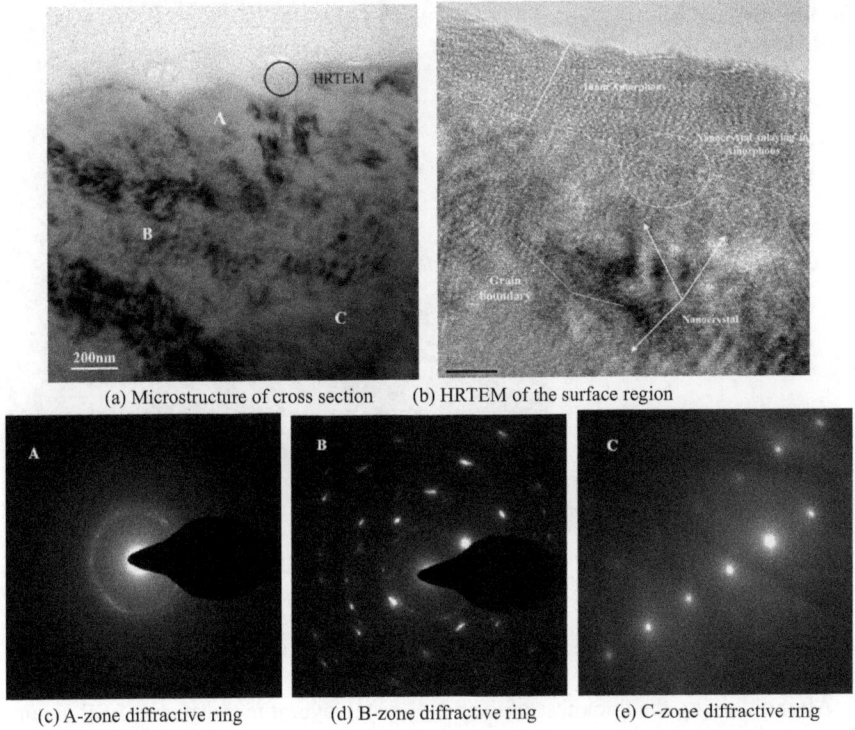

(a) Microstructure of cross section   (b) HRTEM of the surface region

(c) A-zone diffractive ring   (d) B-zone diffractive ring   (e) C-zone diffractive ring

Fig. 3.17 TEM images of cross section of titanium alloy after LSP with a power density of 6 GW/cm^2

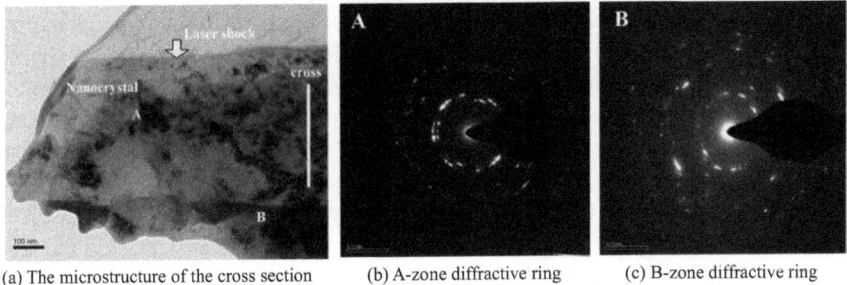

(a) The microstructure of the cross section   (b) A-zone diffractive ring   (c) B-zone diffractive ring

Fig. 3.18 TEM images of a TC6 cross section after 5 impacts with a power density of 6 GW/cm^2

length, and the contrast of grain boundaries is unclear and complex, indicating that the internal stress and lattice distortion are high.

It can be observed from the HRTEM of the surface layer (Fig. 3.19) that an amorphous layer with a thickness of about 10 nm formed on the surface layer after 5 impacts. Compared with other shock parameters, the thickness of the amorphous

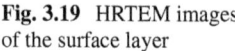

Fig. 3.19 HRTEM images
of the surface layer

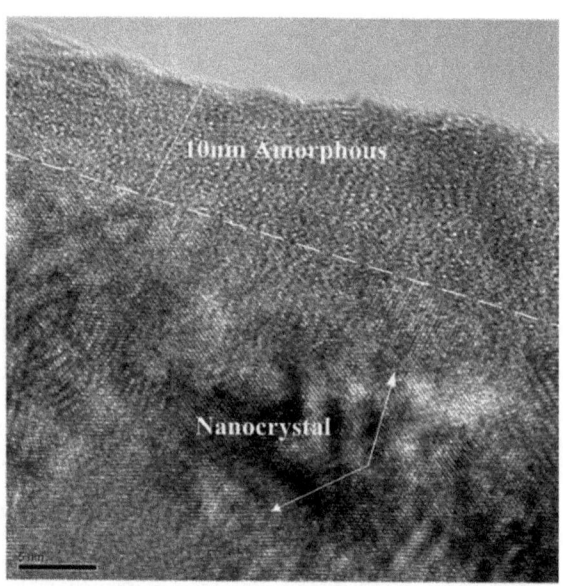

layer has no obvious change, and the nanocrystal layer is distributed in the area below
the amorphous layer.

After obtaining the thickness of the nanocrystal layer under different LSP parameters, the microstructures at different depths were characterized. TEM filmswere
fabricated on the surface layer, 3, 10 μm away from the surface and the substrate
area by two-jet thinning, and can be observed by means of TEM. The results are
shown in Fig. 3.20.

TEM observation films of the internal microstructure of TC6 titanium alloy after
LSP were obtained from the samples subjected to LSP with a power density of
6 GW/cm^2 for 5 times, and only surface nanocrystals are observed for the samples
subjected to LSP with a power density of 6 GW/cm^2 and one impact time, Among
them, Fig. 3.20a–e were taken at the distance of 1.5 mm, 5–20 μm and 3–5 μm
from the impact surface, respectively. It can be seen from the observation of the
TEM that the grain size of TC6 substrate is more than tens of microns, and it is
distributed in two phases (as shown in Fig. 3.20a, b). The microstructures in the area
10 μm away from the surface layer are mainly ultra-high dense dislocations, and
the dislocations piled up at the phase boundary and the grain boundary (Fig. 3.20c).
The microstructures in the area 3 μm away from the surface layer are still mainly
ultra-high dense dislocations, and there are also some very small dislocation cells
(Fig. 3.20d, e).

Nanocrystals with a size of 50–200 nm were formed on the surface layer after
one LSP, but the distribution is not uniform, as shown in Fig. 3.20f, the corresponding diffraction pattern is made up of clear and continuous electron diffraction
rings, indicating that the orientation is random. After a LSP for 5 times, uniform
nanocrystals with the size of 30–100 nm were formed on the surface layer. A

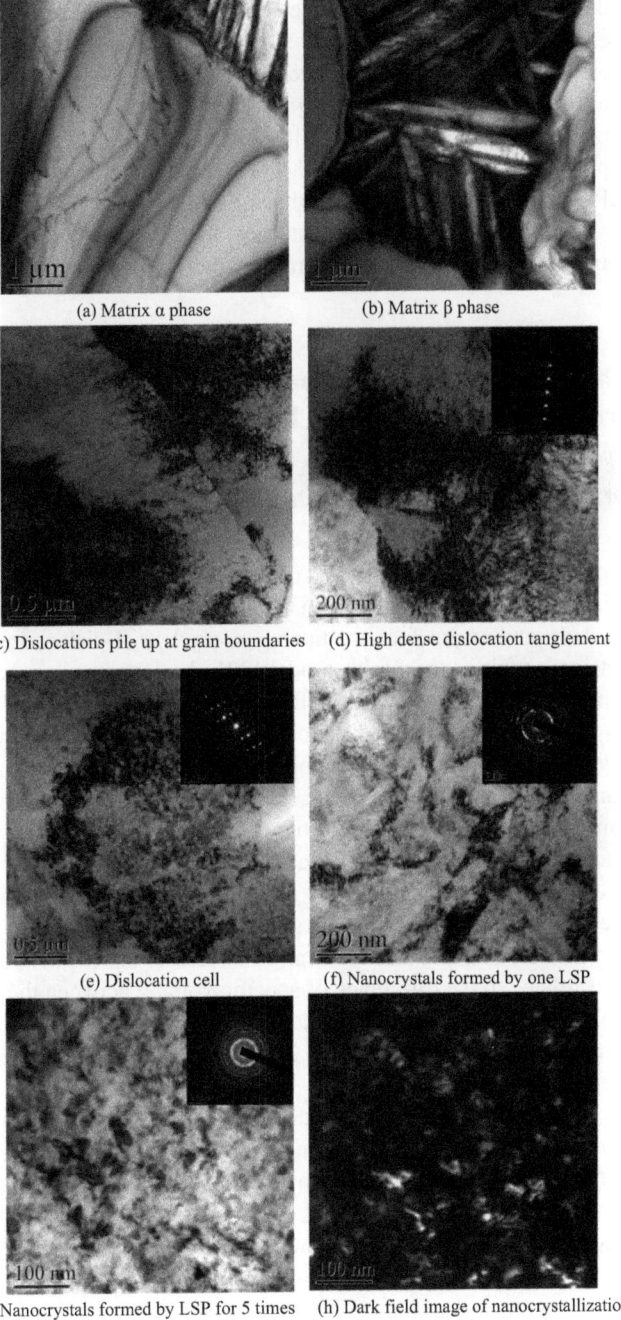

(a) Matrix α phase   (b) Matrix β phase

(c) Dislocations pile up at grain boundaries   (d) High dense dislocation tanglement

(e) Dislocation cell   (f) Nanocrystals formed by one LSP

(g) Nanocrystals formed by LSP for 5 times   (h) Dark field image of nanocrystallization

Fig. 3.20 TEM observation at different depths after LSP for 1 and 5 times

regular circular diffraction pattern indicates random orientation of nanocrystals, as shown in Fig. 3.20g, h. Twins were not found in the TEM observation of the cross-section and plane samples, which indicated that there was no twinning deformation of titanium alloy after LSP. Dislocation movement is the main reason for LSPed nanocrystallization of the TC6 titanium alloy surface.

To sum up, from TEM observation of the TC6 titanium alloy cross section under different LSP parameters, it can be seen that LSP has introduced a gradient structure with an amorphous and nanocrystal surface, high-density dislocations into the subsurface layer and coarse-grained substrate on the TC6 titanium alloy material.

3.4 Mechanism of the Formation of Gradient Microstructure Induced by LSP in Titanium Alloy

3.4.1 Formation of Dislocation in Titanium Alloy by LSP

Among the parameters of shock waves, pressure is the most important one, and the dislocation density will increase with the increase in pressure. Many scholars according to their studies put forward their own models of dislocation generation under impact load, such as the Smith model [22], the Hornbogen model [23], the Mogilevsky model [22], the Weertman-Follansbee model [24] and the Meyers model [23], among which the Meyers model with homogeneous nucleation is widely used.

According to the Meyers shock homogeneous nucleation model (as shown in Fig. 3.21), under the action of a laser shock wave, the deviatoric stress caused by uniaxial strain will homogeneously nucleate dislocation on or near the shock wave front (the nucleation mechanism of dislocation on the shock wave front is unique, which is different from the homogeneous nucleation in conventional deformation), and the dislocation will relax the deviatoric stress. Figure 3.21b shows the wave whose wave front coincides with the first dislocation interface; the dislocation density at the interface is related to the difference of specific volume between the two lattices. As shown in Fig. 3.21c, the wave front moves to the front of the interface, and the deviatoric stress is re-established. In Fig. 3.21d, it can be seen that after elastic waves, new dislocation loops are generated.

When the shock wave is introduced into the material, the high deviatoric stress will distort the initial cubic lattice and form an orthorhombic lattice. When the stress reaches a certain critical value, homogeneous nucleation will occur. Under the shock wave load, the dislocation interface will be separated into two kinds of lattices with different parameters. It can be known from the Meyers dislocation nucleation model that [25]:

$$\tau_h/G = 0.054 \qquad (3.3)$$

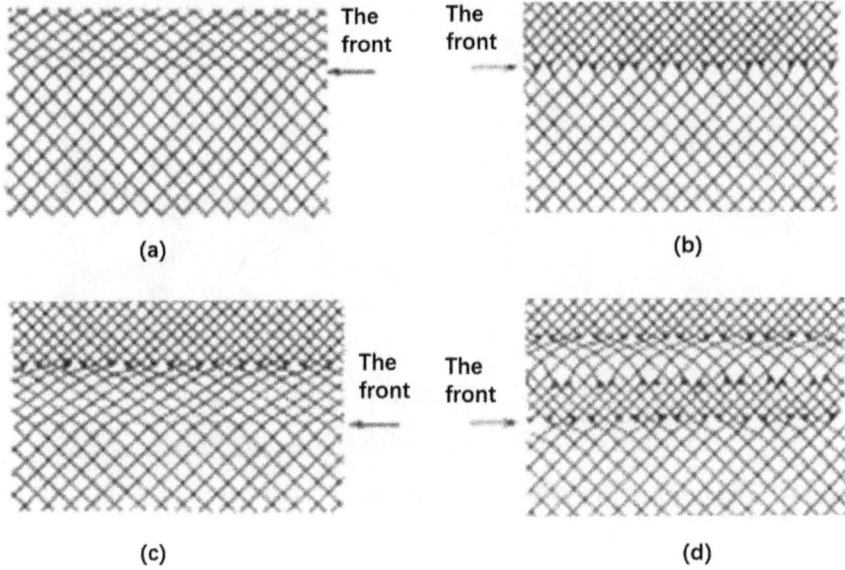

Fig. 3.21 The process of the action of the shock wave front in the Meyers' model

where τ_h is the required shear stress and G is the shear modulus. When the maximum shear stress is equal to τ_h (and acts in the direction which is easy to slip), homogeneous nucleation will occur. The maximum shear stress is about $2GP/K$, where $K = E/3(1-2v)$, the pressure threshold when dislocation nucleates homogeneously is $P = 0.027K$.

For TC6 titanium alloy, its threshold is 2.54 GPa, where s is estimated. Therefore, the peak pressure of the high pressure shock wave induced by the laser can satisfy the condition of homogeneous nucleation of dislocations and form high dense dislocations in the material. In addition, in any case, the distribution of dislocations under impact is more homogeneous than that in conventional deformation. Dhere et al. [26] studied the dislocation substructure of aluminum under impact and cold rolling, in which the distribution of the dislocation of aluminum under impact is more uniform than that under cold rolling.

In the process of shock wave propagation, shock waves may be reflected and transmitted at different interfaces, depending on their properties. There are many interfaces and defects in polycrystalline metal materials, such as grain boundaries, phase boundaries and impurities, which cause reflection and transmission of shock waves, as shown in Fig. 3.22.

When the shock wave propagates from medium A to medium B, the stress wave amplitude produced by reflection (σ_R) and transmission (σ_T) has the following relationship with the incident wave (σ_I):

$$\frac{\sigma_T}{\sigma_I} = \frac{2\rho_B C_B}{\rho_A C_A + \rho_B C_B}, \frac{\sigma_R}{\sigma_I} = \frac{\rho_B C_B - \rho_A C_A}{\rho_A C_A + \rho_B C_B} \qquad (3.4)$$

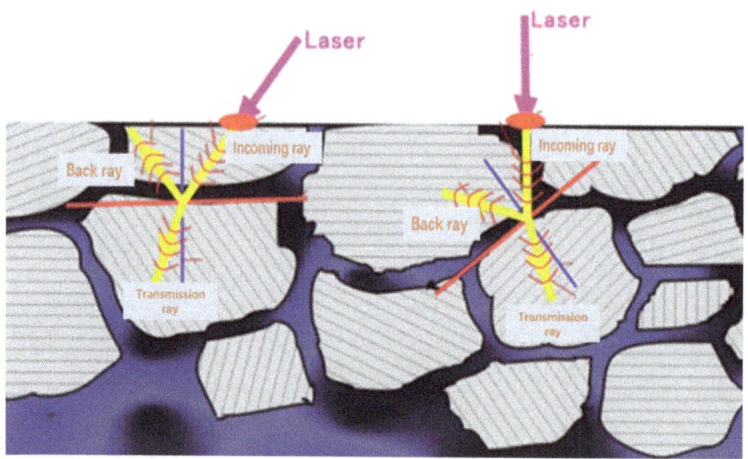

Fig. 3.22 Schematic diagram of reflection and transmission of shock wave on the interface

where C_i is the wave velocity of the elastic wave in medium, and for the uniaxial stress $C_i = \sqrt{E_i/\rho_i}$. For TC6 titanium alloy, the crystal structures of α phase and β phase are different, the elastic modulus and density are also different, and the characteristics of the shock wave propagation are different. Therefore, according to Formula (3.4), the shock wave will reflect and refract at grain boundaries, phase boundaries, impurities and other places, forming a complex wave system. When the reflected and transmitted wave pressure is greater than the dynamic yield strength of the material, it will continue to cause the multiplication and movement of dislocations in the material. Moreover, due to the change in the directions of reflected and transmitted shock waves, slip in different directions will be started to further increase dislocation density. Therefore, dislocation density formed at some interfaces inside titanium alloy and high-temperature alloy is very high, as shown in Fig. 3.23.

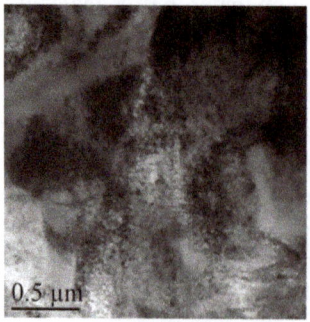
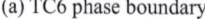
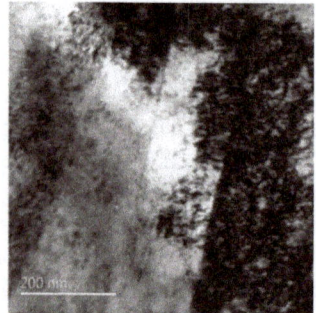

(a) TC6 phase boundary   (b) GH4133B Ni-based superalloy

Fig. 3.23 High dense dislocations at the interface induced by a laser shock wave

3.4.2   Formation of Nanocrystals of the Metal Materials with High Stacking-Fault Energy by LSP

Under the continuous action of shock waves, metal materials continue to undergo strong and severe plastic deformation, and dislocations coordinate through different motion modes. When metals and alloys with high stacking-fault energy undergo plastic deformation, they will quickly form cellular structures, because it is not easy for dislocations in the crystals which have high stacking-fault energy to decompose and have great mobility until they interact with other dislocations to gather and entangle. Therefore, after deformation, the dislocations are unevenly distributed, and the crystal is divided into many areas with high stacking-fault energy and low stacking-fault energy, which is the initial stage of dislocation cells [27]. With the increase in deformation, dislocations on the cell walls interact to form dislocation walls.

The related researchers have proposed a laser shock-wave-induced grain refinement model based on the dislocation model. The model assumes that the dislocation cell structure is formed in the deformation process, which consists of the dislocation of the dislocation cell wall and dislocation inside the dislocation cell. Different dislocation types have different functions: intracellular dislocation density (ρ_c) and dislocation cell wall density (ρ_w), which are divided into two parts: statistical dislocation density (ρ_{ws}) and geometric necessary density (ρ_{wg}). The evolution of these dislocation densities ρ_c, ρ_{ws} and ρ_{wg} along different paths is derived from the following formula:

$$\dot{\rho}_c = \alpha^* \frac{1}{\sqrt{3}b}\sqrt{\rho_{wc} + \rho_{wg}}\dot{\gamma}_w^r - \beta^* \frac{6}{bd(1-f)^{1/3}}\dot{\gamma}_c^r - k_0\left(\frac{\dot{\gamma}_c^r}{\dot{\gamma}_0}\right)^{-1/n}\rho_c\dot{\gamma}_c^r \quad (3.5)$$

$$\dot{\rho}_{ws} = \beta^* \frac{\sqrt{3}(1-f)}{fb}\sqrt{\rho_{ws} + \rho_{wc}}\dot{\gamma}_c^r$$

$$+ (1-\xi)\beta^* \frac{6(1-f)^{2/3}}{bdf}\dot{\gamma}_c^r - k_0\left(\frac{\dot{\gamma}_w^r}{\dot{\gamma}_0}\right)^{-1/n}\rho_{ws}\dot{\gamma}_w^r \quad (3.6)$$

$$\dot{\rho}_{wg} = \xi\beta^* \frac{6(1-f)^{2/3}}{bdf}\dot{\gamma}_c^r \quad (3.7)$$

The first condition of Formulas (3.5) and (3.6) corresponds to the generation of dislocations, the second condition is for the transformation of dislocation cells into dislocation walls, and the last condition is for the dislocations' disappearing during deformation which results in dynamic recovery. Equation (3.7) assumes that the geometric necessary dislocations from the interior of the dislocation cell to the wall of the dislocation cell, α^*, β^* and k_0, are the control parameters of the dislocation evolution rate, n is the dislocation sensitivity parameter, f is the dislocation cell volume fraction, b is dislocation cell size, $\dot{\gamma}_w^r$ and $\dot{\gamma}_c^r$ are cell wall and shear rates of intracellular fracture respectively, and $\dot{\gamma}_0$ is the correlation shear rate. Assuming that the three are equal, the strain compatibility along the interface between the cell and the boundary is satisfied. It can be calculated by $\dot{\gamma}^r = M\dot{\varepsilon}$, where M is the Taylor factor. To obtain the shear stress τ^r, the following formula is adopted:

$$\tau_c^r = \alpha Gb\sqrt{\rho_c}\left(\frac{\dot{\gamma}_c^r}{\dot{\gamma}_0}\right)^{1/m} \tag{3.8}$$

$$\tau_w^r = \alpha Gb\sqrt{\rho_{ws} + \rho_{wg}}\left(\frac{\dot{\gamma}_w^r}{\dot{\gamma}_0}\right)^{1/m} \tag{3.9}$$

$$\tau^r = f\tau_w^r + (1-f)\tau_c^r \tag{3.10}$$

G is the shear modulus, m is the strain rate sensitivity of the material, f is the volume fraction of the dislocation cell wall, and the total dislocation density and dislocation cell size are as follows:

$$f = f_\infty + (f_0 - f_\infty)e \tag{3.11}$$

$$\rho_{tot} = f(\rho_{ws} + \rho_{wg}) + (1-f)\rho_c \tag{3.12}$$

$$d = \frac{K}{\sqrt{\rho_{tot}}} \tag{3.13}$$

The calculation of dislocation density is based on the material model, and the J–C model of ultra-high strain rate deformation is usually used for titanium alloy under an ultra-high strain rate, as follows:

$$\sigma = \left(A_{JC} + B_{JC}\varepsilon^{nJC}\right)\left(1 + C_{al}\ln\left(\frac{\dot{\varepsilon}}{\dot{\varepsilon}_0}\right)\right)\left(1 - \left(\frac{T - T_{ref}}{T_m - T_{ref}}\right)^{mjc}\right) \tag{3.14}$$

The above model is used to calculate the dislocation density and the dislocation cell size of TC6 titanium alloy subjected to LSP. The TC6 performance parameters used in the calculation are: elastic modulus E is 113 GPa, the shear modulus G is 48 GPa, the constant b is 0.256 nm, the density ρ is 4.50 g/cm^3, the thermal expansion coefficient is 9.2, thermal conductivity is 8.0, and the calculation results are shown in Fig. 3.24.

It can be seen from Fig. 3.24 that the shock wave with 6 GPa pressure forms certain ultra-high dense dislocations and dislocation cells inside the material, and the dislocation cell size decreases along the depth direction, and the dislocation cell size is in nanometer order in the area 100 μm away from the impact edge of the material. However, TEM observation of the microstructure of titanium alloy after LSP shows that for the sample subjected to one LSP, dislocation cells are only found

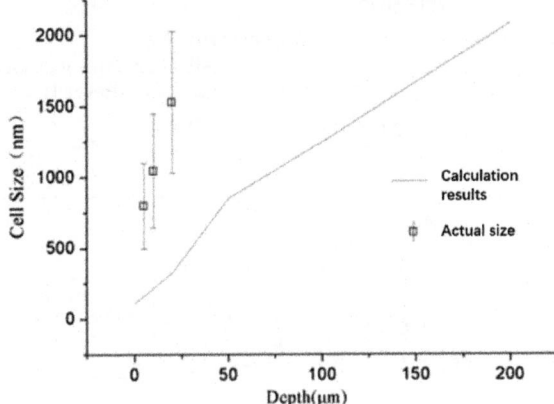

Fig. 3.24 Dislocation cell size in different depth regions under 6GPa shock wave pressure

within 10 μm from the impact edge, and the size of the distribution is very uneven, and the actual size is larger than the calculated size, but it is still in nanometer order.

The strain rate of the laser shock wave is very high, and the rising time of the shock wave is a nanosecond, which provides limited time for dislocation movement. The dislocation multiplication, disappearance and rearrangement are in a small area, and it is easy to form dislocation cells with nanometer spacing. Under the continuous action of a shock wave (or multiple actions), when the dislocation density of the cell wall reaches a certain critical value, the cell wall will be further thinned, and the dislocations of different orientation on the cell wall willinteract with each other, leaving only one redundant dislocation; the dislocation with Burgers vector perpendicular to the cell boundary produces misorientation; when the dislocation density increases, it will cause the transformation from cellular structure to grain structure, as shown in Fig. 3.25.

Combined with the microstructure analysis of synchrotron radiation and TEM observation, the refining process of metal materials with high stacking-fault energy through LSP is summarized as follows: when the shock wave pressure is greater than the homogeneous nucleation threshold of material dislocation, high-density dislocations are rapidly formed near the shock wave front; under the further action of higher pressure and a greater number of times of shock waves, nanocrystals are rapidly formed on the surface of metal materials, and the cross section presents the characteristics of gradient change, as shown in Fig. 3.26.

Fig. 3.25 Dislocation cell evolves into grain boundary

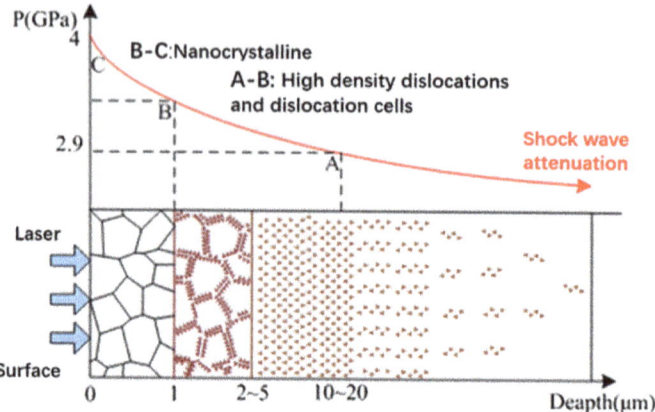

Fig. 3.26 The distribution of the gradient microstructure of high-rise dislocation energy metallic materials by LSP

3.5 The Characteristics of Gradient Microstructures and the Mechanism for the Martensite Transformation of LSP in AISI 304 Stainless Steels

3.5.1 Introduction

AISI 304 stainless steels (AISI 304ss) are widely used when both high strength and good corrosion resistance are required. The high pressure plasma shock wave is induced by laser irradiation on the materials and causes severe plastic deformation in the material [28]. In this work, a surface nanostructured layer was formed on the AISI 304ss after multiple LSP treatments, and the reason for grain refinement is discussed [29].

There are a number of reports for the grain refinement mechanism and martensite transformation of AISI 304ss by severe plastic deformation [30, 31]. For example, according to Lu et al. [20], plastic deformation with an ultra-high strain rate leads to the generation of dislocation lines, dislocation tangles, dislocation walls and mechanical twins (MTs) in the original coarse grains when subjected to multiple LSP. Then the dislocation structures are transformed into subgrain boundaries, which finally evolve into nanoscale grain boundaries through the intersections of mechanical twins and dynamic recrystallization. Chen at el. [32] discussed the deformation mechanism of AISI 304ss subjected to surface impacts over a wide range of strain rates (10–105 s^{-1}) and they found that the strain rate between 10 and 103 s^{-1} only activated dislocation motions and martensite transformations, resulting in nanocrystallines and ultra-fine grains. However, higher strain rates (104–105 s^{-1}) produced a high density of twin bundles with nanoscale thickness in the bulk material. Ye et al. [33] investigated the microstructure evolution of AISI 304ss by LSP at room temperature and at cryogenic temperature (liquid nitrogen temperature) respectively. They found that a nanostructured surface layer was synthesized after LSP at both room temperature and at cryogenic temperature, and indicated that the deformation-induced martensite (DIM) was generated by LSP at room temperature only when the laser-generated plasma pressure was sufficiently high (>5.56GPa). Luo et al. [34] indicated that LSP could not cause deformation-induced martensite. These phenomena may be due to the fact that there is an absorbing layer which avoids the thermal effect from heating the surface by the laser beam during LSP. Gerland and Hallouin [35] investigated the microstructure evolution in AISI 304ss by LSP with a very short laser pulse (0.6 ns) and an extremely high laser intensity (250–1620 GW/cm^2), and they found that the martensite embryos were formed at the intersections of deformation twins within the pressure range of 15–25 GPa. However, these studies do not agree on the conditions for DIM, neither on the role of DIM for grain refinement. The aim of this paper is to reveal the mechanism for the formation of surface nanocrystallines of AISI 304ss induced by multiple LSP treatment by means of Transmission Electron Microscopy observations (TEM), X-ray Diffraction (XRD) and Electron Back Scattered Diffraction (EBSD), and to study the effect of DIM for grain refinement.3.5.1 AISI 304ss and the principle and experimental procedure of LSP.

Samples were cut by a water jet cutter from a plate (thickness 9.0 mm) of AISI 304ss and the sample was annealed in a vacuum condition at 1080 °C for 1 h. Prior to the LSP treatment, the surface of the samples should be polished with SiC paper with the roughness ranging from 500 to 2400 to make the surface roughness at Ra 0.3 μm.

LSP experiments were performed using a Q-switched Nd:YAG laser operating at 1 Hz repetition rate with a wavelength of 1064 nm and the FWHM of the pulses was about 20 ns. The size of the laser beam was 4 mm and the laser intensity used was 4.3 GW/cm^2. Samples were submerged in a water bath, then processed by LSP. A layer of water with a depth of about 1 mm was used as the transparent confining layer and professional aluminum tape with a thickness of 100 μm was used as the absorbing layer. The application of the protective layer is to protect the surface of the metal from direct ablation and to promote a better coupling with the laser energy [36]. The water confinement is to confine the diffusion of the high temperature plasma produced by laser irradiation and to increase the pressure of the shock wave [37].

3.5.2 LSP Induced Surface Nanocrystallization and Martensite Transformation in 304 Stainless Steel

The changes of microstructure and mechanical performance after different LSP impacts are shown in Fig. 3.27. Figure 3.27a, b demonstrate the residual stress within a surface depth of 1600 μm and the Vickers microhardness within a surface depth of 1000 μm of LSPed AISI 304ss samples for one and three impacts. The dotted lines in these two figures indicate the stress and the hardness of the untreated sample, namely,

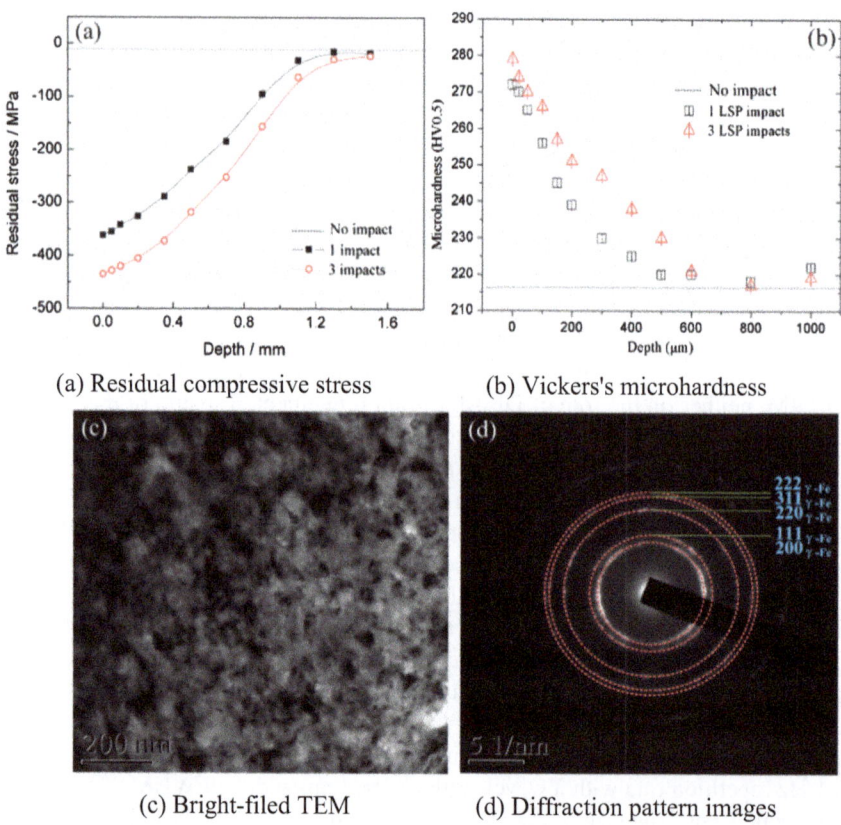

(a) Residual compressive stress   (b) Vickers's microhardness

(c) Bright-filed TEM   (d) Diffraction pattern images

Fig. 3.27 Mechanical properties and microstructure characterization of LSPed AISI 304ss sample

matrix stress and matrix hardness. It is clear that the LSP has a powerful effect on the mechanical properties of the AISI 304ss stainless steel, and a significant increase in residual stress and hardness is observed. Compared to 1 LSP impact, the effect of the 3 LSP impacts is stronger. In addition, the material surface has a maximal residual compressive stress and a maximal microhardness. For example, the sample subjected to 3 LSP has a maximum compressive stress of 435 MPa and a maximum hardness of 279 HV0.5 on the surface, much larger than those of the untreated sample (about 15 MPa and 216 $HV_{0.5}$, respectively). However, an increased depth gives rise to a sharp reduction in the stress and hardness, and they tend to be stable at a certain depth, 1400 and 800 μm, respectively. The enhanced mechanical properties of the LSPed samples can be attributed to the deformation-induced martensite and surface nanocrystallization, as discussed later.

Figure 3.27c shows the TEM image of the surface of AISI 304ss after 3 LSP treatments. It can be observed that the LSP has induced the nanocrystallites with the size of 30–500 nm on the surface of the sample. The corresponding diffraction

Fig. 3.28 XRD patterns of the AISI 304ss surface after LSP at different impacts

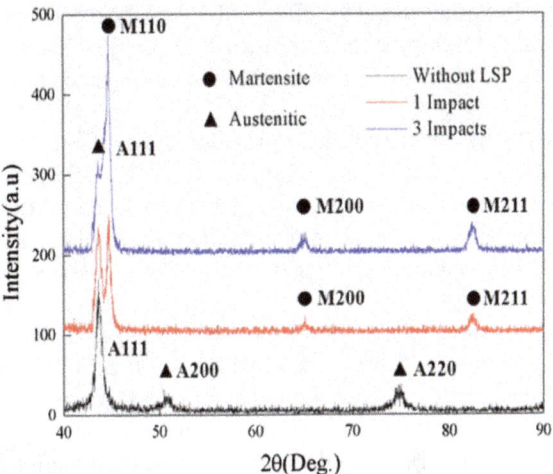

pattern confirms the presence of the nanocrystallites with random orientation, which is dominated by circles, as is shown in Fig. 3.27d.

Much literatures considers that the twinning is a prevalent deformation mechanism of AISI 304ss by severe plastic deformation for the grain refinement [28, 31]. However, besides mechanical twinning, we found that deformation-induced martensite (DIM) is an important deformation mechanism for grain refinement on the top surface at ultra-high strain rates in AISI 304ss subjected to multiple LSP impacts. The XRD qualitative analysis of phases of AISI 304ss treated by different impacts was carried out, as is shown in Fig. 3.28.

Without regard to the influence of instrumental broadening, the Bragg diffraction peak of AISI 304ss have broadened, which indicates that the grain refinement, lattice deformation and the increase in micro-stress have occurred on the surface layer of the alloy. Meanwhile, untreated AISI 304ss only consisted of an austenite phase, but the DIMs took place on the surface layer subjected to different LSP impact treatments. After LSP with a laser intensity of 4.3 GW/cm^2 and a single impact, the two peaks (A200 and A220) corresponding to austenite disappeared, while the three peaks (M110, M200 and M211) corresponding to martensite showed up, indicating that the LSP induced the formation of martensite. In the research of Ye [33], it was also found that the DIM of AISI 304ss subjected to LSP, with a laser power density higher than 5.6 GW/cm^2 was needed. However, in our work, the DIM takes place after a single impact with a laser intensity of 4.3 GW/cm^2. The number of impacts will also greatly affect DIM. After three impacts, three peaks (M110, M200 and M211) were intensified. On the other hand, the peaks (A111) corresponding to the austenite phase decrease in intensity with an increase in the number of impacts, indicating that more and more austenite phases have transformed to martensite upon a greater number of impacts. The purpose of multiple LSP impacts is to provide a longer time and more energy to plastic deformation and thus induce more martensite, and also an increase in the volume fraction of martensite.

To further reveal the effect of DIM on the mechanism of grain refinement, we used EBSD to analyze the microstructure of AISI 304ss surface with and without LSP. The EBSD observations on the surface subjected to three LSP impacts are shown in Fig. 3.29.

In these, two typical deformation-induced microstructural features are identified: MTs and DIM. Figure 3.29a, b shows the EBSD images of the top surface without LSP treatment, which shows that the surface of AISI 304ss is mainly composed of austenite and a small amount of martensite, which is consistent with Chen [32]; this is because martensite transformation was generated with the process of annealing in AISI 304ss.

Figure 3.29c, d shows the EBSD images of the top surface subjected to three LSP impacts. It was found that the MTs and DIMs were induced by multiple LSP impacts. Figure 3.29e, f are magnified images of circle [A] and [B] in Fig. 3.29c, d, showing the typical microstructure observed on the top surface. It can be seen that the martensite was induced by severe plastic deformation and its size was decided by the

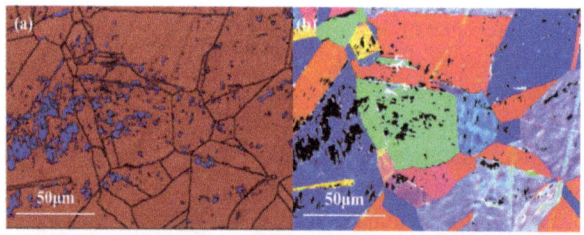

(a) Phase distribution without LSP (b) Grain orientation without LSP

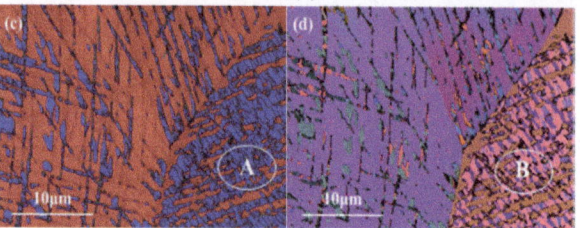

(c) Phase distribution with 3 LSP impacts (d) Grain orientation with 3 LSP impacts

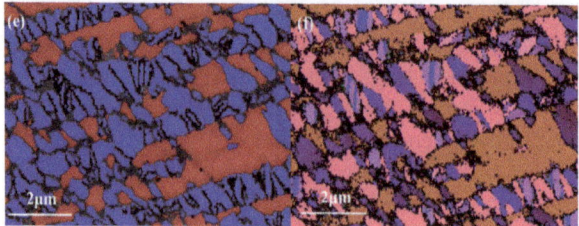

(e) Magnification of local areas A in fig(c)   (f) The magnification of local areas B in fig(d).

Fig. 3.29 EBSD maps of AISI 304ss after 3 LSP impacts; in the phase distribution map, different colors represent different phases, red and blue represent austenite and martensite, respectively. In the grain orientation map, different colors represent different grain orientations

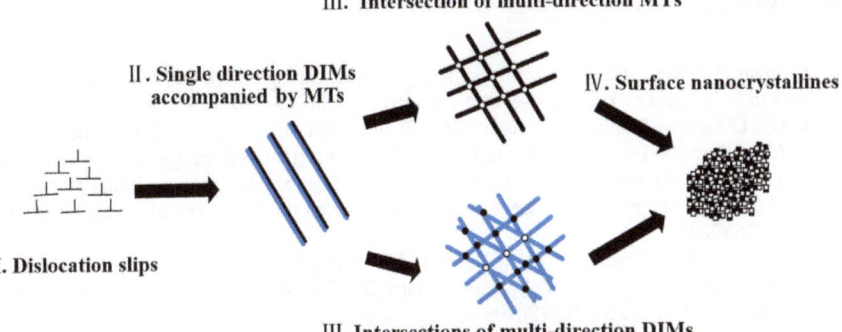

Fig. 3.30 Schematic illustration showing the process of the microstructural evolution of the AISI 304ss

MTs, and DIM at different orientations was intercrossed, and then refined the grains to between 50 and 300 nm. Meanwhile, the lath-shaped martensite was embossed and stratified at different orientations as shown in Fig. 3.29f. Such a phenomenon was induced by the shearing, which was caused by the DIM. Owing to the presence of a habit plane during phase transformation, there was a slight orientation relationship among the impact-formed martensite phases.

The mechanism for the formation of surface nanocrystalline was induced under multiple LSP impacts and can be seen in Fig. 3.30. Each state will be discussed in terms of the experimental observations and literature.

The high dense dislocations are rapidly generated at the wave front by the multi-directional loads which are formed by reflection and refraction of the shock wave [25] as can be seen in Fig. 3.30 state(I).

When the stress caused by the pile-up of dislocation increase to a certain level, MTs will be generated. The boundaries of these MTs are almost parallel, and they divide the original coarse grain into thin twin matrix lamellae on the top surface [38], as can be seen in Fig. 3.30 state (II).

As the wave pressure further increases, the distortion becomes stronger and stronger, finally the accumulated distortion makes the original twins into intercrossed multiple twins [39]. The intercrossed MTs are accompanied by martensite transformation, the size of which was decided by the size of the intercrossed twins. The DIMs were further accompanied by the deformation and intercrossing of twins, as can be seen in Fig. 3.30 state (III).

With the interaction of MTs and DIMs induced by multiple LSP, the coarse grain evolves continually and eventually a nanocrystalline layer with a size of 50–300 nm is formed on the surface of the AISI304ss, as can be seen in Fig. 3.30 state (IV) [40].

References

1. S. Spanrad, J. Tong, Characterization of foreign object damage (FOD) and early fatigue crack growth in laser shock peened Ti-6AL-4V aerofoil specimens. Proc. Eng. (2010)
2. B. Lin, C. Lupton, S. Spanrad, J. Schofield, J. Tong, Fatigue crack growth in laser-shock-peened Ti-6Al-4V aerofoil specimens due to foreign object damage. Int. J. Fatigue **59**, 23–33 (2014)
3. S. Zabeen, M. Preuss, P.J. Withers, Evolution of a laser shock peened residual stress field locally with foreign object damage and subsequent fatigue crack growth. Acta Mater. **83**, 216–226 (2015)
4. X.C. Zhang, Y.K. Zhang, J.Z. Lu, F.Z. Xuan, Z.D. Wang, S.T. Tu, Improvement of fatigue life of Ti-6Al-4V alloy by laser shock peening. Mater. Sci. Eng. A-Struct. Mater. Prop. Microstruct. Process. **527**(15), 3411–3415 (2010)
5. X. Nie, W. He, L. Zhou, Q. Li, X. Wang, Experiment investigation of laser shock peening on TC6 titanium alloy to improve high cycle fatigue performance. Mater. Sci. Eng. A **594**, 161–167 (2014)
6. X. Li, W.F. He, S. Luo, X.F. Nie, L. Tian, X.T. Feng, R.K. Li, Simulation and experimental study on residual stress distribution in titanium alloy treated by laser shock peening with flat-top and Gaussian laser beams. Materials **12**(8) (2019)
7. X. Nie, W. He, S. Zang, X. Wang, J. Zhao, Effect study and application to improve high cycle fatigue resistance of TC11 titanium alloy by laser shock peening with multiple impacts. Surf. Coat. Technol. **253**, 68–75 (2014)
8. A.K. Rai, R. Biswal, R.K. Gupta, R. Singh, S.K. Rai, K. Ranganathan, P. Ganesh, R. Kaul, K.S. Bindra, Study on the effect of multiple laser shock peening on residual stress and microstructural changes in modified 9Cr-1Mo (P91) steel. Surf. Coat. Technol. **358**, 125–135 (2019)
9. S. Prabhakaran, S. Kalainathan, P. Shukla, V.K. Vasudevan, Residual stress, phase, microstructure and mechanical property studies of ultrafine bainitic steel through laser shock peening. Opt. Laser Technol. **115**, 447–458 (2019)
10. Y. Yang, X. Lian, K. Zhou, G. Li, Effects of laser shock peening on microstructures and properties of 2195 Al-Li alloy. J. Alloy. Compd. **781**, 330–336 (2019)
11. X.L. Pan, X. Li, L.C. Zhou, X.T. Feng, S.H. Luo, W.F. He, Effect of residual stress on S-N curves and fracture morphology of Ti6Al4V titanium alloy after laser shock peening without protective coating. Materials **12**(22), 12 (2019)
12. G. Xu, K.Y. Luo, F.Z. Dai, J.Z. Lu, Effects of scanning path and overlapping rate on residual stress of 316L stainless steel blade subjected to massive laser shock peening treatment with square spots. Appl. Surf. Sci. **481**, 1053–1063 (2019)
13. C.Y. Wang, K.Y. Luo, X.Y. Bu, Y.Y. Su, J. Cai, Q.L. Zhang, J.Z. Lu, Laser shock peening-induced surface gradient stress distribution and extension mechanism in corrosion fatigue life of AISI 420 stainless steel. Corrosion Sci. **177** (2020)
14. B. Mao, Y. Liao, B. Li, Gradient twinning microstructure generated by laser shock peening in an AZ31B magnesium alloy. Appl. Surf. Sci. **457**, 342–351 (2018)
15. C.H. Ye, S. Suslov, B.J. Kim, E.A. Stach, G.J. Cheng, Fatigue performance improvement in AISI 4140 steel by dynamic strain aging and dynamic precipitation during warm laser shock peening. Acta Mater. **59**(3), 1014–1025 (2011)
16. X.K. Meng, J.Z. Zhou, C. Su, S. Huang, K.Y. Luo, J. Sheng, W.S. Tan, Residual stress relaxation and its effects on the fatigue properties of Ti6Al4V alloy strengthened by warm laser peening. Mater. Sci. Eng. A-Struct. Mater. Prop. Microstruct. Process. **680**, 297–304 (2017)
17. J.Z. Zhou, X.K. Meng, S. Huang, J. Sheng, J.Z. Lu, Z.R. Yang, C. Su, Effects of warm laser peening at elevated temperature on the low-cycle cross mark fatigue behavior of Ti6Al4V alloy. Mater. Sci. Eng. A-Struct. Mater. Prop. Microstruct. Process. **643**, 86–95 (2015)
18. A. Amanov, R. Karimbaev, E. Maleki, O. Unal, Y.S. Pyun, T. Amanov, Effect of combined shot peening and ultrasonic nanocrystal surface modification processes on the fatigue performance of AISI 304. Surf. Coat. Technol. **358**, 695–705 (2019)

19. M. Benedetti, T. Bortolamedi, V. Fontanari, F. Frendo, Bending fatigue behaviour of differently shot peened Al 6082 T5 alloy. Int. J. Fatigue **26**(8), 889–897 (2004)

20. J.Z. Lu, K.Y. Luo, Y.K. Zhang, G.F. Sun, Y.Y. Gu, J.Z. Zhou, X.D. Ren, X.C. Zhang, L.F. Zhang, K.M. Chen, C.Y. Cui, Y.F. Jiang, A.X. Feng, L. Zhang, Grain refinement mechanism of multiple laser shock processing impacts on ANSI 304 stainless steel. Acta Mater. **58**(16), 5354–5362 (2010)

21. J.Z. Lu, L.J. Wu, G.F. Sun, K.Y. Luo, Y.K. Zhang, J. Cai, C.Y. Cui, X.M. Luo, Microstructural response and grain refinement mechanism of commercially pure titanium subjected to multiple laser shock peening impacts. Acta Mater. **127**, 252–266 (2017)

22. T.L. Cocker, D. Baillie, M. Buruma, L.V. Titova, R.D. Sydora, F. Marsiglio, F.A. Hegmann, Microscopic origin of the Drude-Smith model. Phys. Rev. B **96**(20) (2017)

23. T.C. Germann, D. Tanguy, B.L. Holian, P.S. Lomdahl, M. Mareschal, R. Ravelo, Dislocation structure behind a shock front in Fcc perfect crystals: atomistic simulation results. Metall. Mater. Trans. A-Phys. Metall. Mater. Sci. **35A**(9), 2609–2615 (2004)

24. P.S. Follansbee, J. Weertman, On the question of flow stress at high strain rates controlled by dislocation viscous flow. Mech. Mater. **1**, 345–350 (1982)

25. M.A. Meyers, *Dynamic Behavior of Materials* (Wiley, NewYork, USA, 1994).

26. A.G. Dhere, H.J. Kestenbach, M.A. Meyers, Correlation between texture and substructure of conventionally and shock wave-deformed aluminum. Mater. Sci. Eng. **54**(1), 113–120 (1982)

27. N. Iqbal, N.H. van Dijk, S.E. Offerman, N. Geerlofs, M.P. Moret, L. Katgerman, G.J. Kearley, In situ investigation of the crystallization kinetics and the mechanism of grain refinement in aluminum alloys. Mater. Sci. Eng. A-Struct. Mater. Prop. Microstruct. Process. **416**(1–2), 18–32 (2006)

28. J.P. Cuq-Lelandais, M. Boustie, L. Soulard, L. Berthe, A. Sollier, J. Bontaz-Carion, P. Combis, T. de Resseguier, E. Lescoute, Comparasion between experiments and molecular dynamic simulations of spallation induced by ultra-short laser shock on micrometric tantalum targets, in *Shock Compression of Condensed Matter*, ed. by M.L. Elert, W.T. Buttler, M.D. Furnish, W.W. Anderson, W.G. Proud, Pts 1 and 22009 (2009), pp. 829-+

29. L.C. Zhou, W.F. He, S.H. Luo, C.B. Long, C. Wang, X.F. Nie, G.Y. He, X.J. Shen, Y.H. Li, Laser shock peening induced surface nanocrystallization and martensite transformation in austenitic stainless steel. J. Alloy. Compd. **655**, 66–70 (2016)

30. B.N. Mordyuk, Y.V. Milman, M.O. Iefimov, G.I. Prokopenko, V.V. Silberschmidt, M.I. Danylenko, A.V. Kotko, Characterization of ultrasonically peened and laser-shock peened surface layers of AISI 321 stainless steel. Surf. Coat. Technol. **202**(19), 4875–4883 (2008)

31. H.W. Zhang, Z.K. Hei, G. Liu, J. Lu, K. Lu, Formation of nanostructured surface layer on AISI 304 stainless steel by means of surface mechanical attrition treatment. Acta Mater. **51**(7), 1871–1881 (2003)

32. A.Y. Chen, H.H. Ruan, J. Wang, H.L. Chan, Q. Wang, Q. Li, J. Lu, The influence of strain rate on the microstructure transition of 304 stainless steel. Acta Mater. **59**(9), 3697–3709 (2011)

33. C. Ye, S. Suslov, D. Lin, G.J. Cheng, Deformation-induced martensite and nanotwins by cryogenic laser shock peening of AISI 304 stainless steel and the effects on mechanical properties. Philos. Mag. **92**(11), 1369–1389 (2012)

34. K.Y. Luo, J.Z. Lu, Y.K. Zhang, J.Z. Zhou, L.F. Zhang, F.Z. Dai, L. Zhang, J.W. Zhong, C.Y. Cui, Effects of laser shock processing on mechanical properties and micro-structure of ANSI 304 austenitic stainless steel. Mater. Sci. Eng. A **528**(13), 4783–4788 (2011)

35. M. Gerland, M. Hallouin, Effect of pressure on the microstructure of an austentic strainless-steel shock-loaded by very short laser-pulses. J. Mater. Sci. **29**(2), 345–351 (1994)

36. B.J. Demaske, V.V. Zhakhovsky, N.A. Inogamov, C.T. White, Oleynik, II, Evolution of metastable elastic shockwaves in nickel, in *Shock Compression of Condensed Matter*, ed. by M.L. Elert, W.T. Buttler, J.P. Borg, J.L. Jordan, T.J. Vogler, Pts 1 and 22012 (2011)

37. B. Wu, Y.C. Shin, Laser pulse transmission through the water breakdown plasma in laser shock peening. Appl. Phys. Lett. **88**(4), 041116 (2006)

38. M. Wen, G. Liu, J.-F. Gu, W.-M. Guan, J. Lu, Dislocation evolution in titanium during surface severe plastic deformation. Appl. Surf. Sci. **255**(12), 6097–6102 (2009)

39. X. Wu, N. Tao, Y. Hong, G. Liu, B. Xu, J. Lu, K. Lu, Strain-induced grain refinement of cobalt during surface mechanical attrition treatment. Acta Mater. **53**(3), 681–691 (2005)
40. H.T. Ding, Y.C. Shin, Dislocation density-based modeling of subsurface grain refinement with laser-induced shock compression. Comput. Mater. Sci. **53**(1), 79–88 (2012)

Chapter 4
Improvement of High Cycle Fatigue Performance in the Titanium Alloy by LSP-Induced Gradient Microstructure

4.1 Experiment Method of Vibration Fatigue Test for Titanium Alloy Blade of Aero-Engine

Under the action of periodic airflow excitation force, the vibration fatigue failure caused by blade resonance is the main failure of the aero-engine titanium alloy blade [1]. This book mainly introduces the vibration fatigue tests on the LSPed titanium alloy.

Vibration fatigue is related to the response to structural resonance, so the study of it needs structural dynamics technology, so as to reveal the laws related to the characteristics of structural dynamics. Compared with static fatigue, vibration fatigue has the following characteristics [2–5]: (1) Modal inertia force and damping force occur in vibration fatigue structures, among which the distribution of the damping force is an important factor in determining the magnitude of the response to structural resonance, and structural resonance failure will directly determine the characteristics of the distribution of the strain mode of the modes of response to structural dominant resonance. This is difficult to study in static fatigue analysis, because static fatigue does not consider damping or modal response. (2) In engineering, structural resonance is usually divided into global resonance, component resonance and local resonance. Vibration fatigue is related more to component resonance or local resonance. Some dynamic modal load excitation often causes vibration coupling between local mode and load, while the damaged part is often the one with a larger strain and defect or stress concentration in local resonance. The damage is caused by the joint action of local resonance and stress concentration, which is difficult to study in static fatigue analysis which only considers stress concentration. (3) The characteristics of the crack propagation of vibration fatigue are different from those of static fatigue. For example, the crack propagation towards resonance and leaving resonance must have different velocities. For another example, if the crack propagation makes the resonant frequency of the structure exceed the range of load frequency, the propagation will converge, otherwise it will diverge.

L. Zhou and W. He, *Gradient Microstructure in Laser Shock Peened Materials*, Springer Series in Materials Science 314, https://doi.org/10.1007/978-981-16-1747-8_4

The methods of fatigue testing include the method of group comparison for different test conditions and the up-and-down method for the area of long life [6–8]. The up-and-down method is a small-sample method of testing fatigue life proposed by Dixon and Mood [9], which is used to measure the level of stress under specified fatigue life and is suitable for a long life area. The test starts with a stress level that is higher than the fatigue limit, and then decreases gradually. Test the first specimen plate under the action of stress σ_0. The specimen is damaged before it reaches the specified life, $N = 10^7$, so the second specimen plate is tested under a lower stress σ_1. When the fourth specimen plate is tested, because the specimen plate is not damaged (succeeds) after 10^7 cycles under the action of σ_3, the fifth specimen plate is tested under the higher stress σ_2. Accordingly, if the previous specimen plate is damaged from fewer than 10^7 cycles, the subsequent test will be carried out under a lower level of stress. Where the previous test piece succeeds, the subsequent test will be carried out under a higher stress until all the tests are completed. The difference $\Delta\sigma$ between levels of stress is called "stress increment". During the whole process, the stress increment remains 5% of the predicted fatigue limit.

Test conditions of specimen plates are as follows: The specimen plate is clamped onto one side at room temperature. Fix the specimen plate on the clamp of the vibration table, ensure that the clamping position of each specimen plate is the same, and perform bending vibration at the frequency of the natural vibration of the first-order bending of the specimen plate, as shown in Fig. 4.1.

The sinusoidal signal with a stable frequency sent by the signal generator is amplified by the power amplifier and then drives the electromagnetic vibration table to vibrate at the specified frequency; the counter can accurately count the excitation signals. The tip amplitude is measured by the reading microscope and the eddy

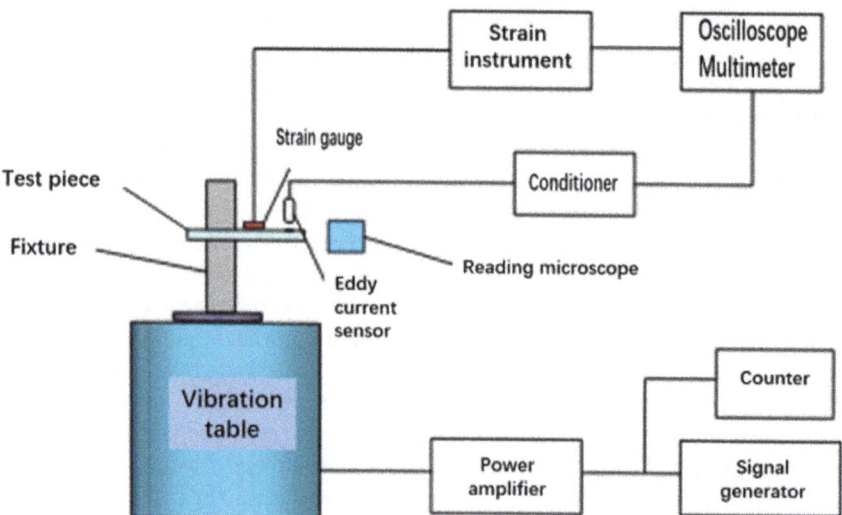

Fig. 4.1 Schematic diagram of experimental system

Fig. 4.2 Clamping mode of specimen on shaking table

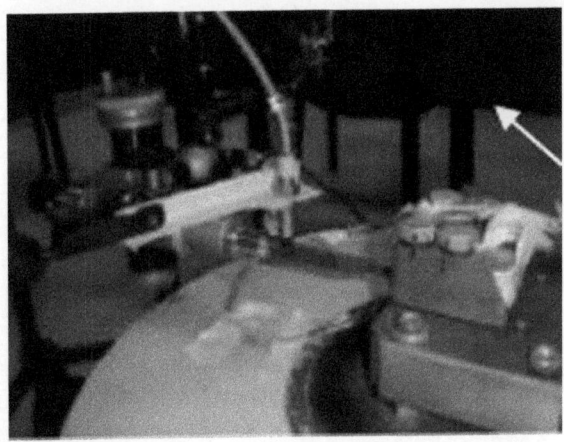

current displacement sensor, the vibration signal measured by the eddy current sensor is conditioned by the conditioner, and then the vibration waveform and amplitude are displayed by the oscilloscope and the multimeter, so the amplitude of the tip of the specimen plate can be monitored conveniently, and the measured value of the eddy current sensor can be calibrated by the reading microscope. Stick a strain gauge in the test area of the specimen plate to measure the stress level during the test. The measured value of the strain gauge is conditioned by the strain gauge and displayed by the oscilloscope and the multimeter; in the experiment, the amplitude of the tip of the specimen plate is controlled by adjusting the frequency and magnitude of the output signal of the signal generator.

A vibration table is used for vibration excitation in the test, which is made up of a specimen excitation system, an amplitude measurement system (reading microscope, eddy current displacement sensor) and a stress measurement system. The composition of the experimental system is shown in Fig. 4.2.

4.2 Improvement of High Cycle Fatigue Performance in the Titanium Alloy by LSP-Induced Gradient Microstructure

4.2.1 Vibration Fatigue Performance of Samples

The vibration fatigue test of TC6 and TC11 titanium alloys at room temperature is conducted. According to the actual heat treatment process of a titanium alloy blade, heat treatment is carried out, and the size and strengthened area of the specimen is shown in Fig. 4.3, and LSP parameters are shown in Table 4.1.

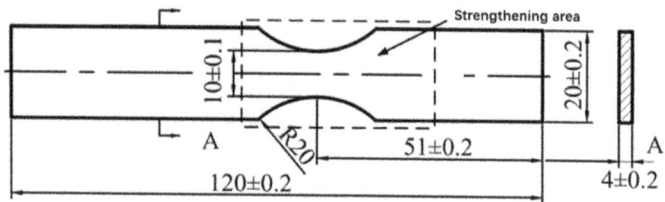

Fig. 4.3 Dimension and strengthening area of standard vibration fatigue specimen

Table 4.1 LSP parameters of TC6 and TC11 titanium alloy sample

Parameter	Laser energy (J)	Spot diameter (mm)	Laser pulse width (ns)	Lap ratio
Value	2–8	3–4	5–20	60–75%

TC6 and TC11 specimens subjected to LSP are shown in Figs. 4.4 and 4.5. The specimens are straight without bending after being processed by double-sided LSP, and double-sided LSP can restrain the deformation of the specimen during impacting. Uniform and continuous smooth micro-pits can be observed in the treated area by the reflection of light, which indicates that the pressure of the plasma shock wave generated by the laser beam exceeds the limit of the dynamic yield of the material, causing plastic deformation. Impact pits have the same size and close arrangement, and there is no overlapping gap between them.

Fig. 4.4 Morphology of TC6 specimen after LSP

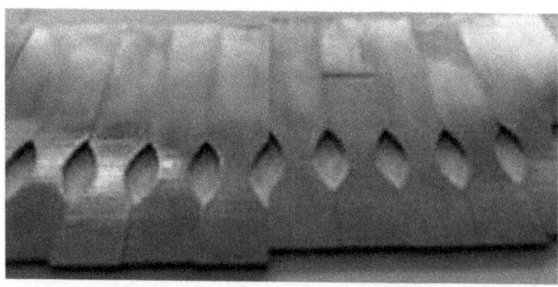

Fig. 4.5 Morphology of TC11 specimen after LSP

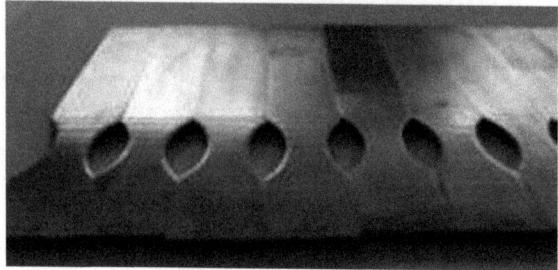

(a) Surface topography

(b) 3D stereogram of surface topography

Fig. 4.6 AFM map of LSPed region

After LSP, the surface roughness of materials will be improved. Under the action of alternating load, the valley of roughness of the surface can easily cause stress concentration and fatigue crack. The greater the value of the surface roughness, the deeper the crack on the surface, the smaller the radius of the bottom of the crack, and the worse the ability to resist fatigue damage; according to the principle of fracture mechanics, the greater the surface roughness, the greater the notch effect, that means the greater the stress concentration coefficient, the worse the fatigue performance.

Figure 4.6 are AFM photos of the TC6 titanium alloy treated by 10 times of LSP. Figure 4.6a shows the contour feature of the treated surface, and the scanning area is 100×100 µm. Figure 4.6b is the 3D stereo photo of the treated area. It can be seen from the figure that the change in surface roughness caused by LSP is small, and the surface roughness Ra after LSP is about 0.46 µm.

The more times LSP is carried out, the greater the depth of micro-pits formed by LSP, and the higher the roughness. However, compared with the traditional surface strengthening technology (such as shot peening), the LSP has less influence on the roughness of the surface of the material, which is 0.46 µm after 10 times of LSP, while the roughness of shot peening with 0.3A is 1.261 µm.

The fatigue strength of treated titanium alloy specimens is tested under 10^7 cycles. Before the formal test, a simulation of calculation of finite elements and a pre-vibration test on the specimens in various states is made to calibrate the relationship between the amplitude of characteristic points and the stress at the assessment site, and determine the initial state and stress increment of the fatigue test. The initial levels of the stress of TC6 and TC11 specimens are 460 MPa and 510 MPa, respectively, and the stress increment is 20 MPa. After LSP, the level of the vibration stress of the two titanium alloys starts at 580 MPa, and the stress increment is constant. See Fig. 4.7 and Fig. 4.8 respectively, for the up-and-down diagram of the vibration fatigue test before and after LSP of the two titanium alloys.

It can be seen from Figs. 4.7 and 4.8 that under the condition of 95% confidence, the original fatigue strength of the TC6 standard vibration fatigue specimen is 439 MPa, and the fatigue strength after LSP is 527 MPa. The fatigue test of the TC6 standard specimen plate shows that the fatigue strength after LSP is improved by 20.1%. The original fatigue strength of the TC11 standard vibration fatigue specimen is 483 MPa, and the fatigue strength after LSP is 593 MPa, increasing by 22.8%. The vibration

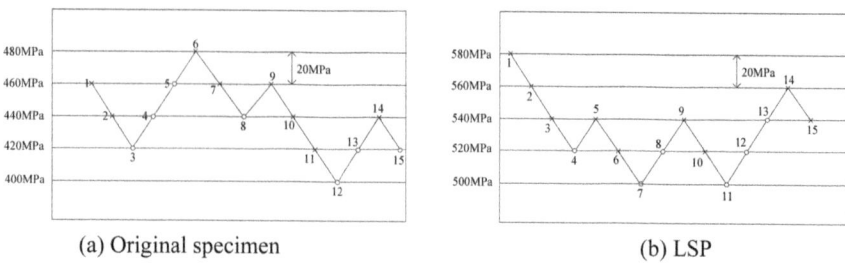

(a) Original specimen   (b) LSP

Fig. 4.7 Lifting diagram of TC6 titanium alloy standard vibration fatigue specimen

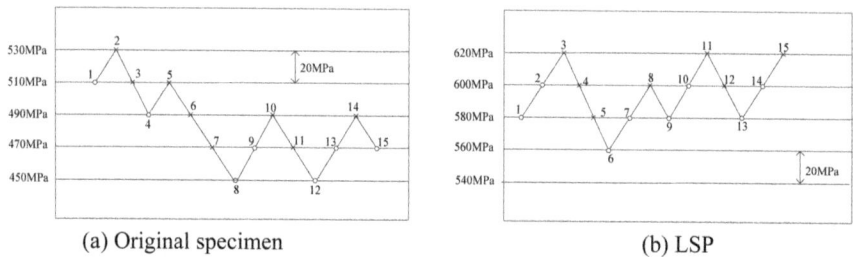

(a) Original specimen   (b) LSP

Fig. 4.8 Lifting diagram of TC11 titanium alloy standard vibration fatigue specimen

fatigue tests of these two titanium alloys before and after LSP show that LSP can effectively improve the fatigue properties of titanium alloys.

The first-order bend fatigue life of the titanium alloy standard vibration fatigue specimen and blade is over 10^5, and its fatigue failure belongs to the category of high cycle fatigue. Continue the fatigue test on the failed specimen until it breaks. See Figs. 4.9 and 4.10 for typical fatigue fracture morphology and local enlargement of the TC11 specimen before and after LSP.

TC11 belongs to a dual-phase titanium alloy, and dense slip bands are formed due to non-uniform deformation caused by anisotropy [10–13]. These slip bands interact with phase boundaries to form many discontinuous damage bands, which gradually form crack sources with the accumulation of damage. The fatigue source of the specimen in the original state is located on the surface of the specimen, as shown in Fig. 4.9a. After LSP, the crack source is located about 0.2 mm in the sub-surface layer of the component, as shown in Fig. 4.10a.

Macro-fractography shows that after the formation of the crack source, the fatigue crack propagates with the crack source as its center and presents two parts: a flat area and an uneven area. The flat area is the characteristic mark of slow and stable crack propagation, and the source of cracks is located in the flat area of fracture; the uneven area is a morphological feature of rapid crack instability and propagation, and there are secondary cracks and fatigue bands perpendicular to the direction of the propagation of the crack in this area [14–16]. Comparing the fracture surfaces before and after LSP treatment, it is found that the flat area of the propagation of the fatigue

(a) Crack source   (b) Slow crack growth region

(c) Fast expanding region and transient fault region   (d) Fatigue bands

Fig. 4.9 Fracture of specimen in original state

crack of the strengthened specimen is obviously larger than that of the untreated specimen. The measurement by CAD software shows that the flat area after LSP is 2.1 times that of the untreated specimen, and there are a large number of cleavage steps in the flat area of the strengthened specimen, which is a feature of fractures in a state of complex stress. Figures 4.9 and 4.10c, d are typical morphologies of the uneven fracture areas before and after LSP treatment; regarding the specimen after LSP treatment, it has fewer secondary cracks and a closer arrangement of fatigue strips.

4.2.2 Rotary Bending Fatigue Performance of Samples at High Temperature

Through Ti60 titanium alloy, the effect of LSP on fatigue performance of titanium alloy at a high temperature is examined. A test of rotary bending at a high temperature is carried out according to the "GBT 4337-2008 Method of Metallic Materials-Fatigue Testing-Rotating Bar Bending". Test data are analyzed and processed according to the "HB/Z112-86 Method of Statistical Analysis for the Fatigue Testing

(a) Crack source   (b) Slow crack growth region

(c) Fast expanding region and transient fault region   (d) Fatigue bands

Fig. 4.10 Fracture surface of TC11 specimen after LSP

of Materials". The shape and size of the specimens are shown in Fig. 4.11, and the heat treatment is carried out according to the actual heat treatment process of an engine blade, and the LSP parameters are shown in Table 4.2.

The rotary bending fatigue life of Ti60 material treated with different LSP parameters is studied by the group comparison method. The fatigue test results are shown in Fig. 4.12. Test conditions: $\sigma_{max} = 450$Mpa, $R = -1$, T $= 500\,°C$.

It can be seen from Fig. 4.12 that the rotary bending fatigue life of titanium alloy (Ti60) specimens at a high temperature after LSP treatment has been improved,

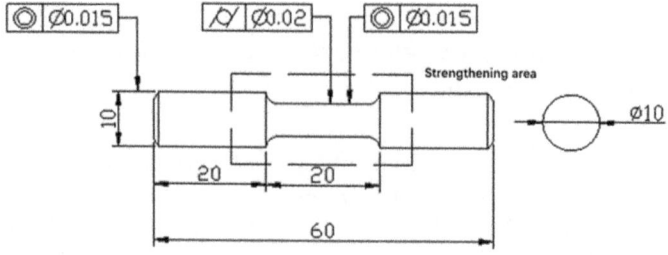

Fig. 4.11 Sample size

Table 4.2 LSP parameters of Ti60 titanium alloy

Parameter	Laser energy (J)	Spot diameter (mm)	Laser pulse width (ns)	Lap ratio
Value	2/4/6/8	3–4	5–20	50%

Fig. 4.12 High temperature rotary bending fatigue life of titanium alloy under different power densities

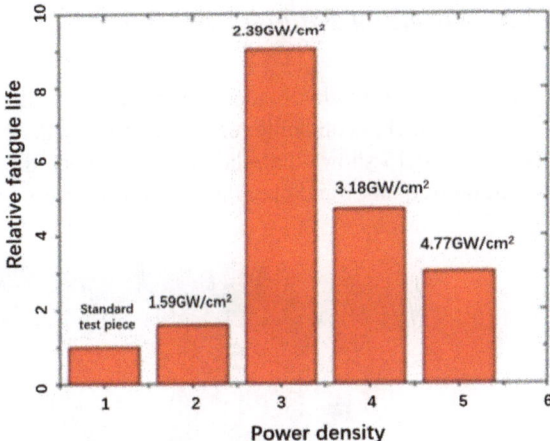

and the fatigue life has the greatest improvement when the power density is 2.39 GW/cm^2. At the same time, when the fatigue test is carried out at 500 °C, the residual compressive stress will be thermally relaxed. Figure 4.13 shows the residual stress relaxation curve of the Ti60 titanium alloy after LSP with a power density of 2.39 GW/cm^2 in high temperature environment. After LSP, residual compressive stress of 650 MPa forms on the surface, and after heat treatment at 500 °C/1 h, the residual stress on the surface relaxes by 50%. Although the residual compressive stress produced by LSP will be released quickly through heat treatment, it can be seen that fatigue performance is still greatly improved through the fatigue experiments.

Fig. 4.13 Surface residual stress relaxation at different holding time

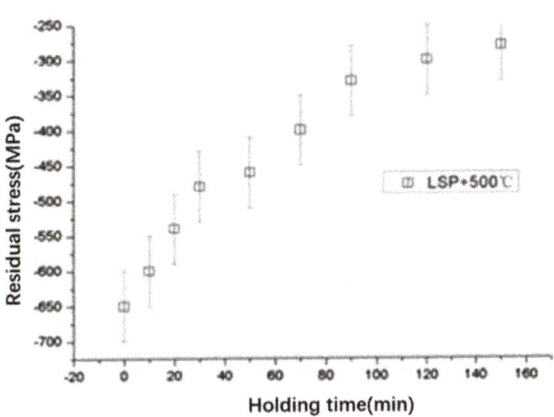

The high temperature fatigue test results of titanium alloy show that: LSP can bring good thermal stability, and the parts of titanium alloy that have been strengthened can be used at 500 °C.

4.2.3 Vibration Fatigue Performance of Blades

In this book, real engine blades are treated by LSP, as shown in Fig. 4.14. In an aero-engine, there is an angle requirement when the blade is equipped with a blade disk. Figure 4.15 shows the relationship between the angle of blade clamping and the natural frequency and maximum vibration stress of a first-order bend mode. The

Fig. 4.14 Blade reinforcement

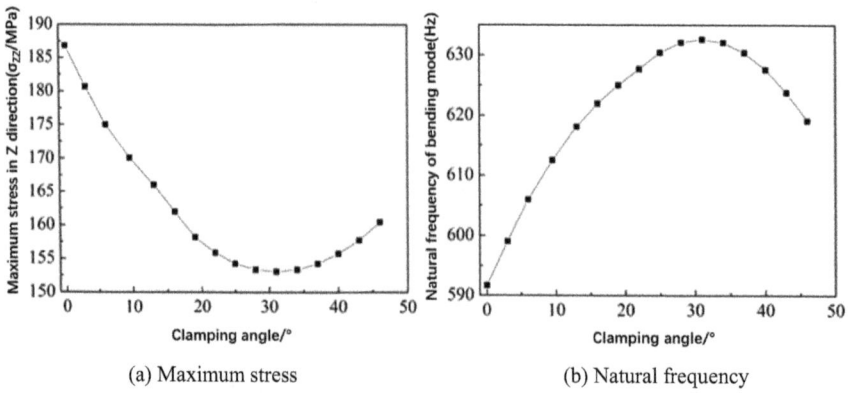

(a) Maximum stress   (b) Natural frequency

Fig. 4.15 Influence of clamping angle on blade mode

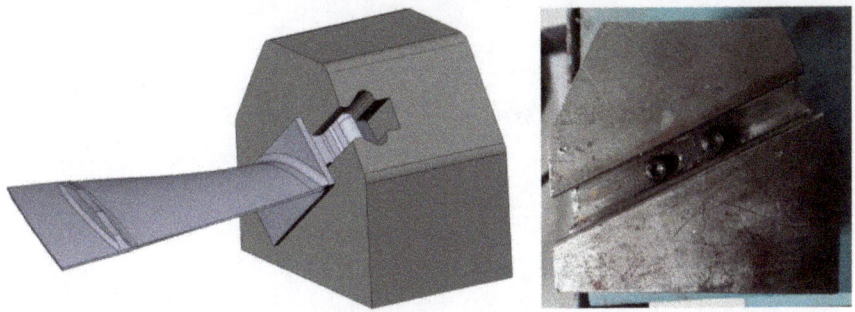

Fig. 4.16 Digital model and real object of vibration fatigue test fixture

numerical simulation results show that the angle of clamping does not change the position of the maximum vibration stress of the first-order bending mode of the blade, but the stress first decreases and then increases with the increase in the angle of clamping, and the natural frequency first increases and then decreases, and they all reach the extreme value when the angle of clamping is 31°.

Therefore, 31° is selected as the angle of clamping of the fatigue test fixture. A bolt hole is preset at the back of the fixture, and the front pressure is preset at the back of the blade tenon during clamping, so that the wall of the tenon is closely attached to the fixture, and the test conditions can be close to the real installation conditions of the blade. See, Fig. 4.16 for clamping and physical objects.

The high cycle fatigue experiment is carried out on a D-300-3 electromagnetic vibration experimental system, and the fixed amplitude of the blade tip and the blade root maximum stress are calibrated by strain gauges. The fatigue strength of the blade before and after LSP is measured by the step-loading method, and the fatigue performance of the blade is evaluated. The test stress cycle number $\triangle N$ is 1×10^7, and the stress increment $\triangle \sigma$ is 30 MPa (about 5% of the initial test load). The high-cycle fatigue test results of the blade are shown in Table 4.3.

Figure 4.17 shows the result of the step-loading fatigue test. The fatigue strength of the original blade measured by step-loading is 481.78 MPa, and the average fatigue strength of the strengthened blade is 541.93 MPa. The fatigue strength of thin titanium

Table 4.3 Vibration fatigue test results

State	Number	Natural frequency (Hz)	Initial stress (MPa)	Loading series ($\times 10^7$)	Total number of cycles	Fatigue limit (MPa)
Original	O-1	654.71	460	1	0.811×10^7	Invalid
	O-2	648.24	430	2	1.726×10^7	481.78
LSP	P-1	639.59	460	3	3.873×10^7	552.19
	P-2	635.36	460	3	1.489×10^7	531.67

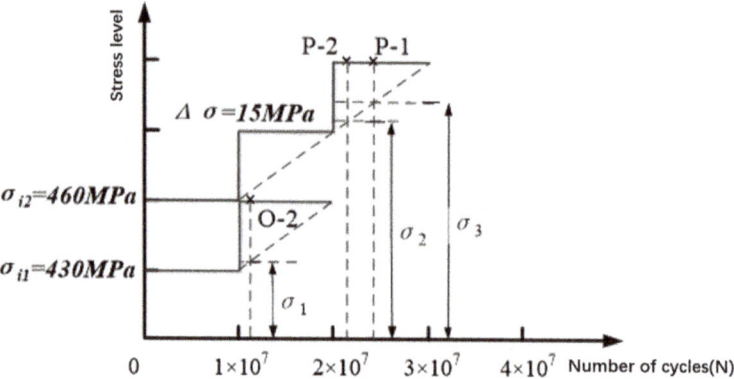

Fig. 4.17 Compressor blade step-loading diagram

alloy blades can be increased by 12.5% by using LSP with certain parameters and processes, which verifies the effectiveness of the parameters and of the process.

The ultrasonic infrared thermal wave nondestructive testing system is used to detect the fatigue crack of blades. The test results are shown in Fig. 4.18. The fatigue crack of the blade in the initial state appears at the maximum stress of the joint line of the first-order bend, while the fatigue crack of the strengthened blade appears at the transition area between the strengthened zone and the non-strengthened area, which is about 8.5 mm away from the joint line of the first-order bend. The analysis shows that LSP can effectively enhance the fatigue resistance of the material in the area of the joint line of the first-order bend of the blade root, and when the amplitude of the blade is increased, the material can effectively avoid failure under high resonance stress. On the other hand, due to the increase in the amplitude, the material stress in the body area of the blade increases, and the transition area becomes a new weak area of the blade under a state of high stress, so the fatigue crack originates here [17, 18].

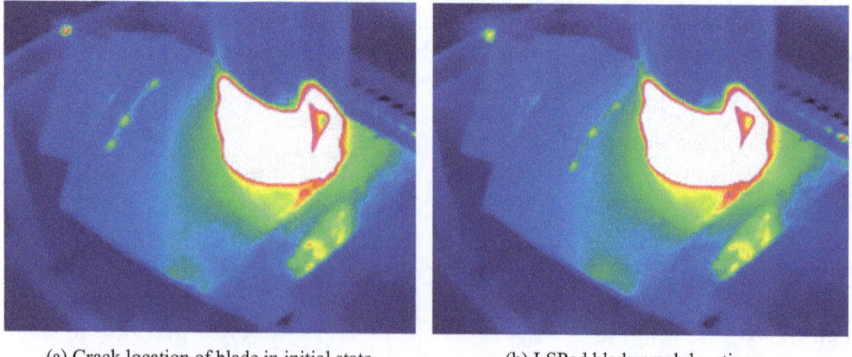

(a) Crack location of blade in initial state   (b) LSPed blade crack location

Fig. 4.18 Ultrasonic infrared thermal wave nondestructive testing of fatigue cracks

4.3 High Cycle Fatigue Performance in LSPed Titanium Alloys Subjected to Foreign Object Damage

4.3.1 Test of Titanium Alloy Blades Injured by Foreign Objects in Air Cannon

The resistance of titanium alloy blades to foreign objects is an important technical index for the evaluation of the structural strength of key components of an aero-engine. In this section, based on the design of the foreign object damage test of titanium alloy blades, the titanium alloy blade is strengthened by LSP treatment, and the effect of LSP on improving the performance of anti-foreign object damage of titanium alloy blades is assessed. In the test, rectangular plate specimens are used, and their shapes and sizes are shown in Fig. 4.19.

The method of high-speed ballistic impact is used to simulate the blade's foreign object damage. The air cannon high-speed ballistic impact test system is used in the test, as shown in Fig. 4.20. The device is mainly made up of an air chamber, a barrel, a velocimeter and protectors. How it works: Nitrogen enters the air chamber through the air inlet valve. When the pressure in the air chamber reaches the preset pressure, the air inlet is stopped and the firing electromagnetic valve is triggered, so that high-pressure nitrogen enters the barrel to push and accelerate the projectile, which then hits the projectile holder separator to separate the projectile holder from the projectile, the projectile holder is blocked in front of the separator, then the projectile impacts the specimen to form damage.

The speed of the projectile can be adjusted by the pressure in air chamber, and the impact velocity is measured by a laser velocimeter. By adjusting the fixed angle of the specimen, the angle of the impact of a projectile on the specimen can be controlled. In order to improve the acceleration of the projectile, a lightweight plastic support is used to hold the projectile, as shown in Fig. 4.21. The use of the projectile holder can also make it so that the shape and size of the projectile are not limited by the barrel.

There are 30 specimens in total, including 15 untreated samples and 15 strengthened samples. The impact speed is 200 and 300 m/s, and the impact angle is 90°. The diameter of steel ball manufactured by GCr15 steel materials is 2 mm. The impact angle and position are shown in Fig. 4.22.

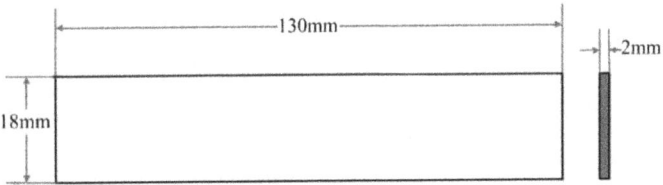

Fig. 4.19 TC4 titanium alloy plate specimen

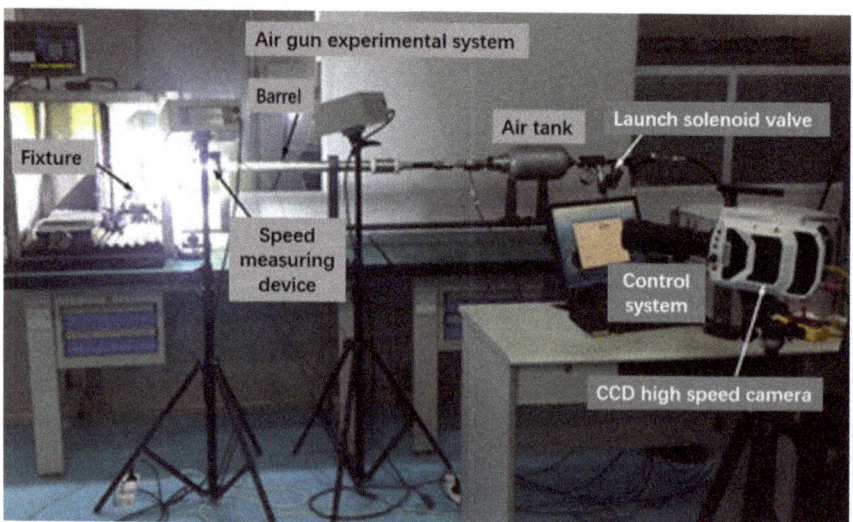

Fig. 4.20 FOD test device

Fig. 4.21 Sabot and steel ball

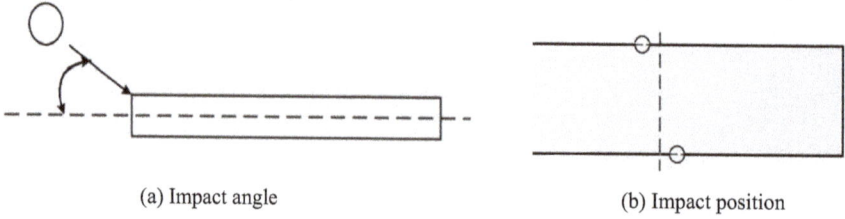

(a) Impact angle   (b) Impact position

Fig. 4.22 Schematic diagram of impact angle and position

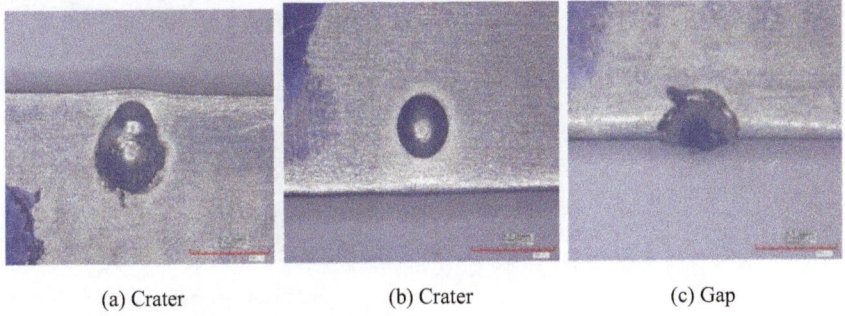

(a) Crater   (b) Crater   (c) Gap

Fig. 4.23 Typical FOD damage morphology

Due to the dispersion of the test, the damage made by the impact is mainly divided into two forms: notches or pits. Figure 4.23 is the typical morphology of the damage of the FOD test. Figure 4.23a shows the positions of impact close to the edge of the specimen, where the projectiles form pits, but bulges occur at the edge of the specimen; Fig. 4.23b shows the positions of impact far away from the edge of the specimen, where they form the typical pit damage; Fig. 4.23c shows the ideal position of impact, where they cause notch damage.

Figures 4.24 and 4.25 show the morphology of the notch caused by projectiles in 300 m/s. Due to the extrusion of projectiles, severe plastic deformation occurred on the upper surface of the notch, and materials piled up and folded along the edge of the notch, forming a fold belt. The fold belt swells and spreads to the outer edge of the notch. Due to the high-speed impact of steel balls, the material splashes near the point of impact. An obvious adiabatic shear band is formed on the upper surface of the notch (because the time of impact is very short, 10^{-7} s, it can be considered as an adiabatic process), as shown in Fig. 4.24a. The formation of the adiabatic shear band is caused by local deformation of materials under the load of impact [19–21]. This local deformation under a high strain rate can cause an obvious rise in temperature in this area; once the material softening caused by a rise in temperature exceeds the material hardening caused by deformation and strain rate effect, this local deformation will develop in a positive feedback way, thus forming an adiabatic shear band.

The periphery of the notch is smooth and flat, which is formed by the initial high-speed penetration and extrusion of a steel ball, showing typical characteristics of compression upon impact [22–24]. At the bottom of the notch, due to the shearing and punching action of a steel ball, material loss occurs, and obvious micro-notch caused by material loss is formed. The shear failure area of the untreated specimen runs through the bottom of the notch, while the shear failure area of the strengthened specimen is smaller.

Macroscopically, the damage notch is mainly divided into three typical areas, namely adiabatic shear area, smooth extrusion area and material loss area, as shown in Fig. 4.26.

Fig. 4.24 Untreated
specimen

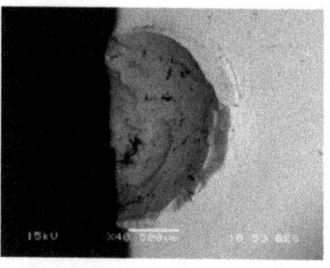

(a) Macroscopic morphology

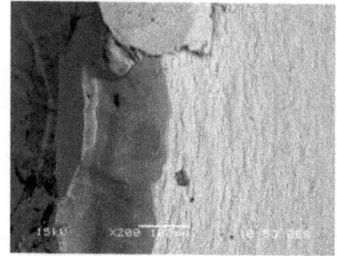

(b) Morphology of edge folds on notch

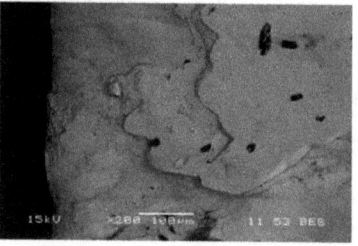

(c) Morphology of material loss at the bottom of the notch

4.3.2 Fatigue Behavior of TC4 Titanium Alloys Damaged by a Foreign Object Under LSP

High-cycle fatigue test of titanium alloy simulated blades damaged by foreign objects is carried out by a pull-pull fatigue test. The fatigue test adopts the step-loading method. During the experiment, the fatigue strength of the sample is tested by the step-loading method based on the linear cumulative damage theory proposed by Maxwell and Nicholas [25, 26]. The stress cycle ΔN of each stage is 1×10^6 cycles, and the stress ratio is 0.1. If no fatigue fracture occurs in the first 1×10^6 cycles, the stress will be increased by $\Delta \sigma$ (10% of the initial stress) to continue the test; if the specimen is still unbroken, continue to increase the stress level, and the stress level should be increased by a 3–4 order. If the cycle times at the first stress level fail to reach 1×10^6, the stress level shall be reduced to restart the test. The stress level at

Fig. 4.25 LSPed specimen

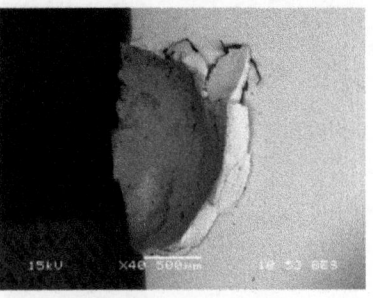

(a) Macroscopic morphology

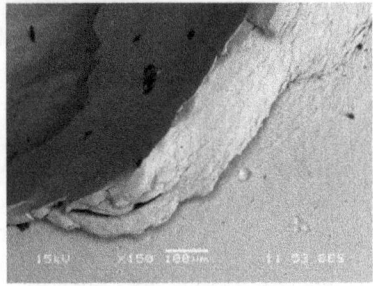

(b) Morphology of edge folds on notch

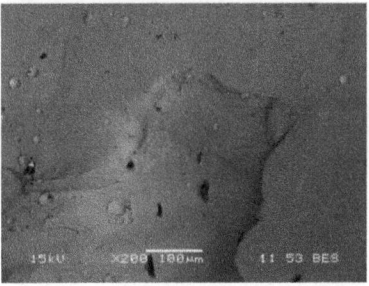

(c) Morphology of material loss at the bottom of the notch

the last failure is σ_l, the stress level before the last failure is σ_{lb}, and the cycle times at the last stress level σ_l are N_l, so the fatigue limit σ of the specimen shall be:

$$\sigma = \sigma_{lb} + \frac{N_l}{\Delta N}\Delta\sigma \tag{4.1}$$

In this test, three kinds of samples are designed for a comparative fatigue test. Among them, sample 1 is a smooth specimen, which has been subjected to LSP and was not damaged by foreign objects; sample 2 is a titanium alloy simulated blade damaged by 200 m/s steel balls; sample 3 is a titanium alloy sample, which was first

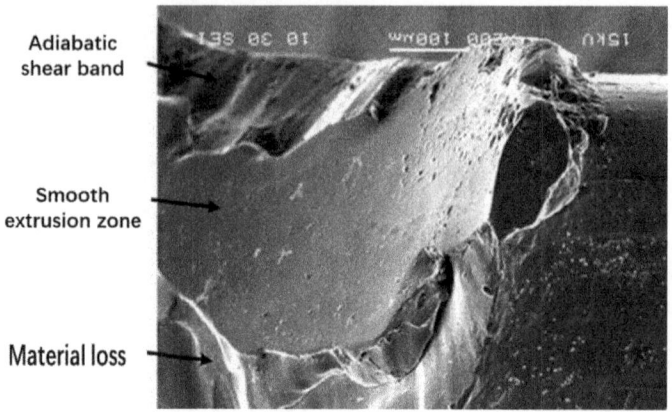

Fig. 4.26 Typical damage morphology

subejected to LSP, and then damaged by 200 m/s steel balls. In addition, due to the complexity of damage by a foreign object and the dispersion of FOD test, the damage caused by the test is quite different. Samples damaged by foreign objects with high consistency are selected for examination, and 10 samples of each type are selected for comparative test. The results are shown in Tables 4.4, 4.5 and 4.6.

After processing the test data in Tables 4.4, 4.5 and 4.6, it can be seen that the fatigue strength of the TC4 blade is 409.8 MPa, the fatigue strength of the TC4 blade after FOD is 274.9 MPa, and the fatigue strength of the LSPed TC4 blade after FOD is 331.9 MPa. It can be seen that the existence of a notch will reduce the fatigue strength of the sample. Compared with the original titanium alloy blade sample, its strength decreases by about 33.1%, and the main reason for the decrease in fatigue performance is the large tensile stress and concentration of the stress at the bottom of the notch. After LSP, the fatigue strength of the sample after being damaged

Table 4.4 Fatigue test results of smooth samples

Number	Initial stress (MPa)	Loading series	Fatigue limit (MPa)
F-1	290	5	415.6
F-2	290	4	399.1
F-3	290	5	418.5
F-4	290	4	397.5
F-5	290	4	385.6
F-6	290	4	392.1
F-7	290	5	432.9
F-8	290	5	422.7
F-9	290	4	409.2
F-10	290	5	425.1

Table 4.5 Fatigue test results of FOD samples

Number	Initial stress (MPa)	Loading series	Fatigue limit (MPa)
F-1	200	4	298.7
F-2	200	3	267.4
F-3	200	1	224.7
F-4	200	3	273.8
F-5	200	5	334.6
F-6	200	4	308.6
F-7	200	2	245.3
F-8	200	2	235.4
F-9	200	4	291.3
F-10	200	3	269.5

Table 4.6 Fatigue test results of LSP + FOD specimens

Number	Initial stress (MPa)	Loading series	Fatigue limit (MPa)
LF-1	260	3	335.8
LF-2	260	2	308.7
LF-3	260	3	332.9
LF-4	260	4	356.7
LF-5	260	3	331.5
LF-6	260	4	376.4
LF-7	260	2	291.5
LF-8	260	2	292.8
LF-9	260	4	351.4
LF-10	260	3	341.5

by foreign objects increased by 20%. This is because LSP can introduce residual compressive stress on the surface of the sample, and the generation of compressive stress can inhibit the propagation of cracks and improve its fatigue strength.

4.3.3 Improving the Tolerance to Damage of the Titanium Alloy Blade by LSP

Literature shows that LSP can not only improve the fatigue performance of damaged titanium alloy blades, but also improve the tolerance to damage of titanium alloy blades [27–30]. Therefore, in this section, the prefabricated notched titanium alloy blades with different notch depths are selected as the research objects, and some of the samples were treated by LSP, and then the fatigue experiments are compared to verify the influence of LSP on the improvement of the tolerance to damage by foreign objects of titanium alloy blades.

Two notched TC4 titanium alloy samples as shown in Fig. 4.27 were designed for the investigation. The length, width and thickness of the sample with a small notch and a large notch were the same, and they were 150 mm, 20 mm and 2 mm respectively. Among them, the small notch was a notch with a depth of 0.1 mm and a width of 0.2 mm cut by wire cutting at the center line position, and the large notch was a notch with a depth of 1.5 mm and a width of 3 mm cut at the center line position. Some samples with large notches were subjected to LSP before cutting for the notch. Parameters of LSP are shown in Table 4.7. The high-cycle fatigue test was carried out by using the step-loading method, in which the stress ratio was $R = 0.1$,

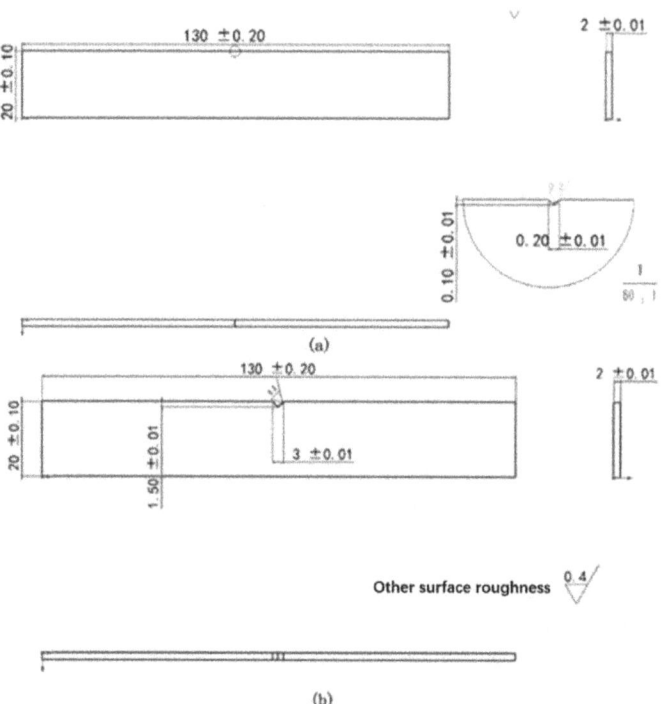

Fig. 4.27 Design of a notched specimen

Table 4.7 LSP parameters

Material	Wavelength λ (nm)	Energy E (J)	Pulse width τ (ns)	Spot size d (mm)	Lap ratio (%)	Power density I (GW/cm²)
TC4	1064	4	20	2.2	50	4.3

Table 4.8 Fatigue test results of small notch specimen

Number	Initial stress (MPa)	Loading series	Fatigue limit (MPa)
C1-1	230	4	334.3
C1-2	230	2	269.5
C1-3	230	2	272.5
C1-4	230	4	321.9
C1-5	230	3	338.9
C1-6	230	3	297.2
C1-7	230	4	335.7
C1-8	230	4	320.6
C1-9	230	3	301.8
C1-10	230	2	269.4

the number of cycles was $\Delta N = 10^6$, and the loading amount of each stage was 5% of the initial stress. The results are shown in Tables 4.8, 4.9 and 4.10.

By processing the test data in Tables 4.8, 4.9 and 4.10 in the same way as calculating the fatigue strength of smooth samples, the fatigue strengths of the sample with a small notch (C1), the sample with a large notch (C2) and the LSPed sample with a large notch (LC2) are 306.1 MPa, 265.5 MPa and 303.9 MPa respectively. The fatigue strength of the smooth TC4 blade is 409.8 MPa. It can be seen that the existence of a notch will greatly reduce the fatigue strength of the sample, and the main reason for the decrease in fatigue performance is the great tensile stress and the concentration of stress at the bottom of the notch. In addition, the greater the notch depth, the more obvious the decrease in fatigue strength. However, the fatigue strength of the LSPed sample with a large notch is similar to that of the sample with

Table 4.9 Fatigue test results of large notch specimen

Number	Initial stress (MPa)	Loading series	Fatigue limit (MPa)
C2-1	230	2	265.7
C2-2	230	1	233.4
C2-3	230	2	272.2
C2-4	230	3	299.7
C2-5	230	2	276.5
C2-6	230	2	264.2
C2-7	230	1	249.2
C2-8	230	2	287.5
C2-9	230	2	274.8
C2-10	230	1	232.1

Table 4.10 Fatigue test results of LSP + large notch specimen

Number	Initial stress (MPa)	Loading series	Fatigue limit (MPa)
LC2-1	230	3	302.1
LC2-2	230	2	289.4
LC2-3	230	3	308.7
LC2-4	230	2	288.2
LC2-5	230	4	333.4
LC2-6	230	2	268.5
LC2-7	230	3	315.2
LC2-8	230	3	311.4
LC2-9	230	4	342.6
LC2-10	230	2	279.3

a small notch. This is because LSP can introduce residual compressive stress on the surface of the sample, and the generation of compressive stress can inhibit the propagation of cracks and improve its fatigue strength. To a certain extent, it can be explained that the depth of the bearable notch for no fatigue failure of LSPed material is more than 15 times better than that of no LSPed material, that is, LSP can increase the tolerance to damage of the material by more than 15 times.

Furthermore, the influence of LSP on the tolerance to foreign object damage caused by air cannon of simulated leading edge blades was also studied in this section. The test samples were designed as shown in Fig. 4.28. The leading edge of the test sample is bent to avoid the influence of neutral layer considering the need of bending fatigue test. The leading edge of half of the samples were treated by LSP with a double-sided LSP process before the foreign object damage test and the parameters

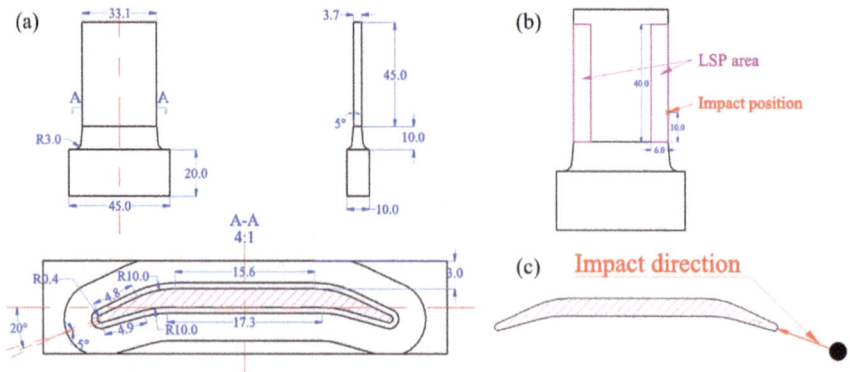

Fig. 4.28 Schematic diagram of simulation blade design and the position of LSP and FOD

are shown in Table 4.7. The air cannon method was used to carry out the FOD test. The test samples without LSP treatment were impacted by 2-mm-diameter sand of 150 m/s and the test samples treated by LSP were impacted by 2-mm-diameter steel ball of 300 m/s. Considering the dispersion of FOD tests conducted by the air cannon method, 5 samples with a notch depth of about 0.1 mm caused by sand and 5 samples with a notch depth of about 1.5 mm caused by steel balls were selected for the high cycle fatigue test. The fatigue tests were carried out on a vibrating table using the step-loading method. The stress ratio is -1, the stress cycle of each step is 10^7, and the stress increment of each step remains 10% of the initial stress. The test results are shown in Tables 4.11 and 4.12.

The relationship between the fatigue limit and the damage size is shown in Fig. 4.29. The average fatigue limit of untreated samples with a 0.1 mm notch impacted by sand is 251.42 MPa, and the average fatigue limit of LSPed samples with a 1.5 mm notch impacted by steel ball is 246.43 MPa, which is similar as the former. It can be seen from the figure that the fatigue limit of the LSPed sample with a 0.11 mm notch is similar to that of the sample with a 1.72 mm. Therefore, the same conclusions can be drawn as in Tables 4.8, 4.9 and 4.10, LSP can increase the tolerance to foreign object damage of the leading edge blade by about 16.2 times. The main reason is that the residual compressive stress introduced by LSP at the leading edge of the blade has a better inhibitory effect on crack propagation and therefore improved fatigue strength.

Table 4.11 Fatigue test results of specimen impacted by sand

Specimen number	Notch depth/mm	Initial stress/MPa	Loading series	Fatigue limit/MPa
S-1	0.122	200	3	244.08
S-2	0.107	200	3	258.96
S-3	0.128	200	2	234.12
S-4	0.093	200	4	277.32
S-5	0.110	200	3	242.66

Table 4.12 Fatigue test results of LSPed specimen impacted by steel ball

Specimen number	Notch depth/mm	Initial stress/MPa	Loading series	Fatigue limit/MPa
LSP-1	1.532	200	3	252.88
LSP-2	1.511	200	4	265.47
LSP-3	1.723	200	2	223.15
LSP-4	1.479	200	3	250.12
LSP-5	1.630	200	2	240.53

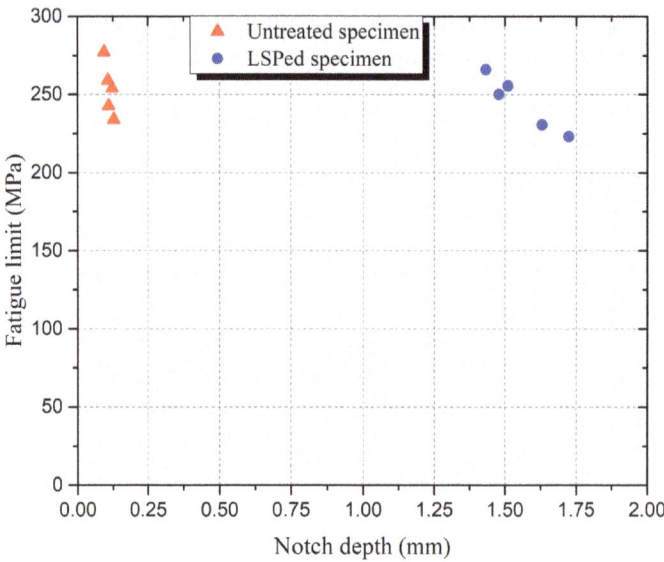

Fig. 4.29 HCF test results

References

1. X. Nie, W. He, S. Zang, X. Wang, J. Zhao, Effect study and application to improve high cycle fatigue resistance of TC11 titanium alloy by laser shock peening with multiple impacts. Surf. Coat. Technol. **253**, 68–75 (2014)
2. H. Mayer, Fatigue crack growth and threshold measurements at very high frequencies. Int. Mater. Rev. **44**(1), 1–34 (1999)
3. A.D. Dimarogonas, Vibration of cracked structures: a state of the art review. Eng. Fract. Mech. **55**(5), 831–857 (1996)
4. R.B. Waterhouse, Fretting fatigue. Int. Mater. Rev. **37**(2), 77–97 (1992)
5. P.D. McFadden, detecting fatigue cracks in gears by amplitude and phase demodulation of the meshing vibration. J. Vibr. Acoust. Stress Reliab. Des. Trans. ASME **108**(2), 165–170 (1986)
6. R. Pollak, A. Palazotto, T. Nicholas, A simulation-based investigation of the staircase method for fatigue strength testing. Mech. Mater. **38**(12), 1170–1181 (2006)
7. Y. Kim, H.J. Lee, D.N. Little, Y.R. Kim, Aapt, A simple testing method to evaluate fatigue fracture and damage performance of asphalt mixtures, in *2006 Journal of the Association of Asphalt Paving Technologists: From the Proceedings of the Technical Sessions*, vol 75 (2006), pp. 755–787
8. S.K. Lin, Y.L. Lee, M.W. Lu, Evaluation of the staircase and the accelerated test methods for fatigue limit distributions. Int. J. Fatigue **23**(1), 75–83 (2001)
9. W.E. Kappauf, An empirical sampling study of the Dixon and mood statistics for the up-and-down method of sensitivity testing. Am. J. Psychol. **82**(1), 40–55 (1969)
10. Q. Xue, M.A. Meyers, V.F. Nesterenko, Self-organization of shear bands in titanium and Ti-6Al-4V alloy. Acta Mater. **50**(3), 575–596 (2002)
11. L. Wagner, Mechanical surface treatments on titanium, aluminum and magnesium alloys. Mater. Sci. Eng. A-Struct. Mater. Prop. Microstruct. Process. **263**(2), 210–216 (1999)
12. S. Suri, G.B. Viswanathan, T. Neeraj, D.H. Hou, M.J. Mills, Room temperature deformation and mechanisms of slip transmission in oriented single-colony crystals of an alpha/beta titanium alloy. Acta Mater. **47**(3), 1019–1034 (1999)

13. E.O. Ezugwu, Z.M. Wang, Titanium alloys and their machinability—a review. J. Mater. Process. Technol. **68**(3), 262–274 (1997)

14. H.K. Rafi, T.L. Starr, B.E. Stucker, A comparison of the tensile, fatigue, and fracture behavior of Ti-6Al-4V and 15–5 PH stainless steel parts made by selective laser melting. Int. J. Adv. Manuf. Technol. **69**(5–8), 1299–1309 (2013)

15. K. Tokaji, M. Kamakura, Y. Ishiizumi, N. Hasegawa, Fatigue behaviour and fracture mechanism of a rolled AZ31 magnesium alloy. Int. J. Fatigue **26**(11), 1217–1224 (2004)

16. C. Rubio-Gonzalez, J.L. Ocana, G. Gomez-Rosas, C. Molpeceres, M. Paredes, A. Banderas, J. Porro, M. Morales, Effect of laser shock processing on fatigue crack growth and fracture toughness of 6061–T6 aluminum alloy. Mater. Sci. Eng. A-Struct. Mater. Prop. Microstruct. Process. **386**(1–2), 291–295 (2004)

17. Z. Mazur, A. Luna-Ramirez, J.A. Juarez-Islas, A. Campos-Amezcua, Failure analysis of a gas turbine blade made of Inconel 738LC alloy. Eng. Fail. Anal. **12**(3), 474–486 (2005)

18. J.F. Hou, B.J. Wicks, R.A. Antoniou, An investigation of fatigue failures of turbine blades in a gas turbine engine by mechanical analysis. Eng. Fail. Anal. **9**(2), 201–211 (2002)

19. S.R. Choi, Foreign object damage phenomenon by steel ball projectiles in a SiC/SiC ceramic matrix composite at ambient and elevated temperatures. J. Am. Ceram. Soc. **91**(9), 2963–2968 (2008)

20. X. Chen, R. Wang, N. Yao, A.G. Evans, J.W. Hutchinson, R.W. Bruce, Foreign object damage in a thermal barrier system: mechanisms and simulations. Mater. Sci. Eng. A-Struct. Mater. Prop. Microstruct. Process. **352**(1–2), 221–231 (2003)

21. J.O. Peters, R.O. Ritchie, Influence of foreign-object damage on crack initiation and early crack growth during high-cycle fatigue of Ti-6Al-4V. Eng. Fract. Mech. **67**(3), 193–207 (2000)

22. S. Zabeen, M. Preuss, P.J. Withers, Evolution of a laser shock peened residual stress field locally with foreign object damage and subsequent fatigue crack growth. Acta Mater. **83**, 216–226 (2015)

23. S. Spanrad, J. Tong, Characterisation of foreign object damage (FOD) and early fatigue crack growth in laser shock peened Ti-6Al-4V aerofoil specimens. Mater. Sci. Eng. A-Struct. Mater. Prop. Microstruct. Process. **528**(4–5), 2128–2136 (2011)

24. J.O. Peters, O. Roder, B.L. Boyce, A.W. Thompson, R.O. Ritchie, Role of foreign-object damage on thresholds for high-cycle fatigue in Ti-6Al-4V. Metall. Mater. Trans. A-Phys. Metall. Mater. Sci. **31**(6), 1571–1583 (2000)

25. T. Nicholas, D.C. Maxwell, Mean Stress Effects on the High Cycle Fatigue Limit Stress in Ti-6Al-4V, Fatigue Fracture Mechanics 33(Dec) (2002) p.476–492.

26. D. Lanning, G.K. Haritos, T. Nicholas, D.C. Maxwell, Low-cycle fatigue/high-cycle fatigue interactions in notched Ti-6Al-4V. Fatigue Fract. Eng. Mater. Struct. **24**(9), 565–577 (2001)

27. Y. Yang, W.F. Zhou, B.Q. Chen, Z.P. Tong, L. Chen, X.D. Ren, Fatigue behaviors of foreign object damaged Ti-6Al-4V alloys under laser shock peening. Int. J. Fatigue **136** (2020)

28. J.F. Wu, Z.G. Che, S.K. Zou, Z.W. Cao, R.J. Sun, Surface integrity of TA19 notched simulated blades with laser shock peening and its effect on fatigue strength. J. Mater. Eng. Perform. **29**(8), 5184–5194 (2020)

29. X.D. Ren, B.Q. Chen, J.F. Jiao, Y. Yang, W.F. Zhou, Z.P. Tong, Fatigue behavior of double-sided laser shock peened Ti-6Al-4V thin blade subjected to foreign object damage. Opt. Laser Technol. **121** (2020)

30. S.H. Luo, X.F. Nie, L.C. Zhou, Y.M. Li, W.F. He, High cycle fatigue performance in laser shock peened TC4 titanium alloys subjected to foreign object damage. J. Mater. Eng. Perform. **27**(3), 1466–1474 (2018)

Chapter 5
Improvement of High Temperature Fatigue Performance in Ni-Based Alloys by LSP-Induced Gradient Microstructures

5.1 Introduction

Ni-based superalloys, the main material used for aero-engine turbine blades [1, 2], are always served in the presence of cyclic loading, especially thermal loading. There is an interest in extending the application of LSP to the aircraft engine parts which are exposed to high temperatures, where the temperatures can reach up to 800 °C [3, 4].

According to quantities of reports on impact simulations and experiments [5–8], the enhancement of the fatigue strength of materials caused by LSP is largely due to the existence of residual compressive stress [5–9]. However, the residual stress shows relaxation when LSP is applied to high-temperature components of aircraft engines due to the thermal effect. Zhou et al. [9] investigated the thermal relaxation behavior of the LSP-induced residual stress at different temperatures between 550 and 700 °C using experimental and finite element simulation methods and found that the residual compressive stress (about 700 MPa on the surface) and the depth of residual stress (~0.4–0.6 mm) are obtained after LSP treatment. Li et al. [10] found that 72% of the residual stress was relaxed after heat treatment of 900 °C/150 min. Zhou et al. [11] investigated the thermal relaxation of residual stress induced by LSP in the K417 alloy at temperatures ranging from 500 to 900 °C using an experimental method and found that residual stress releases rapidly at different temperatures in the initial 30 min and at higher temperatures, it results in faster stress relaxation. Similar results were reported by Nikitin et al. [12] and Ren et al. [13]. All in all, the relaxation of residual stress weakened the strengthening effect of LSP on fatigue performance.

However, LSP will not only form high amplitude residual compressive stress on the surface of the material, but it would also introduce gradient microstructures into the material [9, 14–17]. It leads to grain refinement [9–11], dynamic recrystallization [13, 18] and phase transformation [19] on the surface of metallic materials,

© Zhejiang University Press 2021
L. Zhou and W. He, *Gradient Microstructure in Laser Shock Peened Materials*, Springer Series in Materials Science 314, https://doi.org/10.1007/978-981-16-1747-8_5

which improves their mechanical properties and fatigue performance. Microstructural changes exhibit good thermal stability under thermal action or mechanical loading compared with inductive residual stresses. Huang et al. [18] investigated the microstructure evolution, phase transformation and thermal stability of the nanograined layer of the Ti-25Nb-3Mo-3Zr-2Sn titanium alloy during a 300–600 °C annealing treatment. It was found that the activation energy of grain growth is much lower than that of diffusion in coarse grain. The abnormal low value is caused by the grain boundaries of the surface grains of the alloy not being in equilibrium. In addition, Altenberger et al. [19] found that the improvement in fatigue resistance of the Ti-6Al-4 V alloy at an elevated temperature is related to the high-temperature stability of the highly tangled and dense dislocation substructure induced by LSP. Jia et al. [20] found that the dislocation cells have a good thermal stability in a near α titanium alloy after LSP treatment. However, there are relatively few studies devoted to the determination of the thermal stability of surface nanocrystallines produced by LSP in Ni-based alloys.

In this section, the gradient microstructure was fabricated on the surface of GH4133B and K417 Ni-based superalloys by means of LSP, and the mechanism of thermal stability of the gradient microstructure was discussed in detail. Furthermore, the physical mechanism of the formation of the gradient microstructure induced by LSP was revealed.

5.2 Gradient Microstructure Characteristics Induced by LSP in the GH4133B Ni-Based Superalloy

5.2.1 GH4133B Ni-Based Superalloy and the Principle and Experimental Procedure of LSP

A Ni-based superalloy, named GH4133B, was used for experimental measurements, which is widely employed as a turbine-blade material in aero engines. Table 5.1 lists the mass fractions of chemical compositions in GH4133B.

Commonly, fatigue cracks occur at the tenon teeth positions of turbine blades, especially at the R-transfer between the first tenon tooth and the stretch root. This

Table 5.1 Mass fractions (%) of chemical compositions in GH4133B

Cr	Ti	Al	Fe	C	B	Cu
19.0–22.0	2.65–3.00	0.75–1.15	≤1.5	≤0.06	≤0.01	≤0.07
P	S	Pb	Sn	Sb	As	Bi
≤0.015	≤0.007	≤0.010	≤0.0012	≤0.0025	≤0.0025	≤0.001
Mn	Ce	Mg	Si	Nb	Zr	Ni
≤0.35	≤0.01	0.001–0.01	≤0.60	1.30–1.70	0.01–0.05	Other

Fig. 5.1 GH4133B alloy
turbine blade and LSP area

Table 5.2 LSP parameters

Parameters	Value
Laser wavelength (nm)	1064
Pulse energy (J)	10
Pulse duration (ns)	20
Spot diameter (mm)	3
Repetition rate (Hz)	1
System ASE energy (mJ)	<50
Export laser energy stability	<±5%
Lapping rate	60%

area is close to the first order flexural vibration nodal line of the blades, and is also in contact with the turbine discs. Therefore, the stress at this area is relatively high while working, where the fatigue cracks often occur.

According to the geometrical characteristics of the tenon teeth of a turbine blade, the LSP area is designed, as shown in Fig. 5.1. Turbine blades were submerged in a water bath, and were then processed by LSP. A 1-mm-thick water layer was used as the transparent confining layer, and 100-μm-thick professional aluminum tape was used as the absorbing layer, which protected the sample surface from thermal effects [21]. The detailed parameters used in LSP experiment is shown in Table 5.2.

5.2.2 Gradient Microstructure Induced by LSP and Its Thermal Stability

XRD patterns of surface layers of the samples without and with LSP are shown in Fig. 5.2. The GH4133B alloy is a dendritic branch structure and a γ-Ni (major phase) and γ'-NiAl$_3$ (strengthening phase) solid solution. It can be observed that LSP

Fig. 5.2 XRD patterns of
surface microstructures of
the GH4133B Ni-based
superalloy before and after
LSP

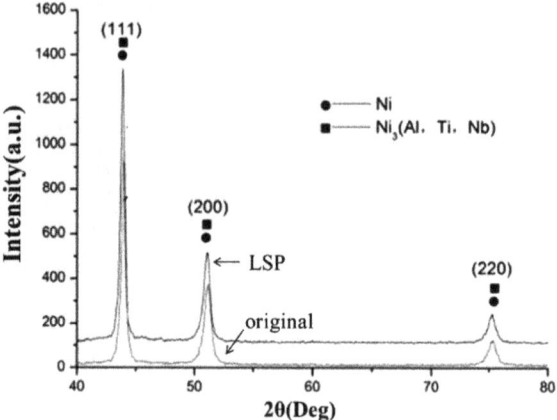

treatment induces no movement in the diffraction peaks, indicating that the structure
of the GH4133B alloy remains unchanged and no new phase is generated.

Then the cross-sectional TEM observation was carried out on the sample subjected
to single LSP with power density of 7.64 GW/cm^2, as shown in Fig. 5.3, in which
the upper side is the LSPedsurface of the sample and protected by C deposition.

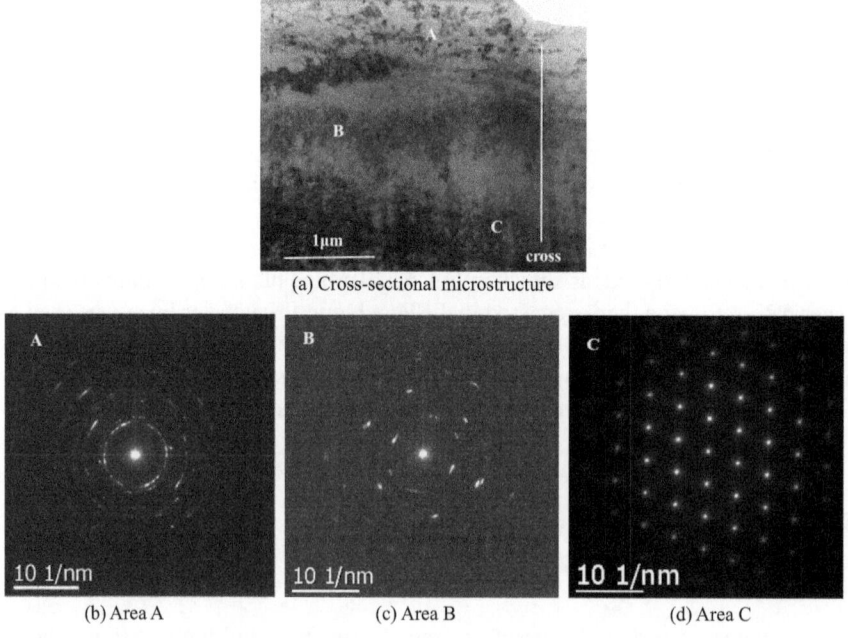

(a) Cross-sectional microstructure

(b) Area A   (c) Area B   (d) Area C

Fig. 5.3 TEM observation of cross section of a GH4133B Ni-based superalloy after a single LSP

It can be seen from the figure that the LSP produced nanocrystals on the surface area, and the thickness of the nanocrystal layer is 0.5–1 μm, and the electron diffraction on the surface area (area A) is round (as shown in Fig. 5.3b), which confirmed the existence of surface nanocrystals with random orientation [22–24]. There are high dense dislocations and elongated grains under the nanocrystalline layer, and no phase boundary was found; the lattice pattern of the transition area (area B) is elongated (Fig. 5.3c), indicating that there is a big lattice distortion here; in the substrate area (area C) about 5 μm away from the surface, a high dense dislocation structure can still be seen, and the corresponding lattice pattern (Fig. 5.3d) shows electron diffraction spots of the typical FCC lattice selected area, indicating that the crystal structure in this area has not changed greatly.

In order to observe the distribution of surface nanocrystals, the image of the area is enlarged locally, as shown in Fig. 5.4. It can be observed that nanocrystals are formed on the surface of the material by LSP, and some grains are elongated perpendicular to the shock direction. The minor axis size is 30–100 nm and the major axis size is about 50–500 nm. Due to the shear stress of the laser shock wave, it is easy to form elongated dislocation cells, and further form elongated nanocrystals.

Without changing its power density, after increasing the impacting number to 3, themicrostructure of the cross section can be observed as shown in Fig. 5.4. The left side of the figure shows the LSPed surface. In order to obtain the original surface structure, C deposition was carried out on the surface.

It can be seen from the figure that nanocrystals were formed on the surface of the GH4133B Ni-based superalloy after the 3 3 times of LSP, with a thickness of 0.5–1 μm, and high-density dislocation tangles and dislocation cells are distributed under the nanocrystal layer, and the size of the dislocation cell is in nanometers. Comparing Fig. 5.3 with Fig. 5.4, it can be seen that increasing the number of impact times does not significantly increase the affected depth of the nano-layer, but the depth of the

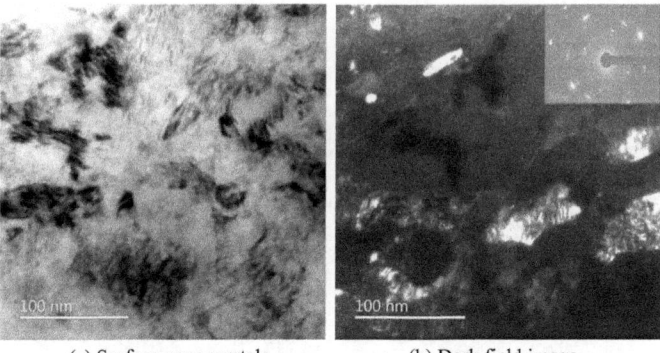

(a) Surface nanocrystals   (b) Dark field image

Fig. 5.4   Surface nanocrystals formed by a single LSP

Fig. 5.5 Cross-sectional
microstructure after 3 times
of LSP

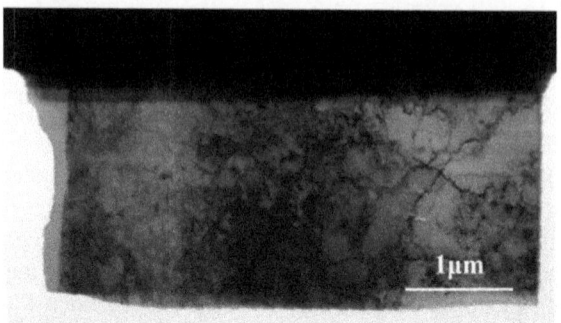

plastic deformation layer increases, which can be reflected in the formation of high-density dislocations in deeper areas on TEM images. Enlarged views of two random areas of the surface layer are shown in Fig. 5.6.

It can be seen from the enlarged view of the local area of the surface that evenly distributed nanocrystals are formed on the surface layer after 3 times of LSP, and the size of elongated grains are further reduced and become equiaxed nanocrystals after increasing the number of impact times. Through the observation of the microstructure at different depths, the typical microstructure 2–5 μm away from the surface is found as shown in Fig. 5.7. It can be observed that the high-strain-rate plastic deformation induced by LSP forms a high-density dislocation structure in the Ni-based superalloy GH4133B, and the density is extremely high, as shown in Fig. 5.7a. Under the further action of a shock wave, dislocations will interact and undergo rearrangement and entanglement to form a large number of dislocation cell structures, as shown in Fig. 5.7b. References [25–29] show that high-pressure shock waves can produce high-density dislocations in metal materials, and the density is extremely high, which is much higher than the dislocation density formed by mechanical treatment, such as shot peening.

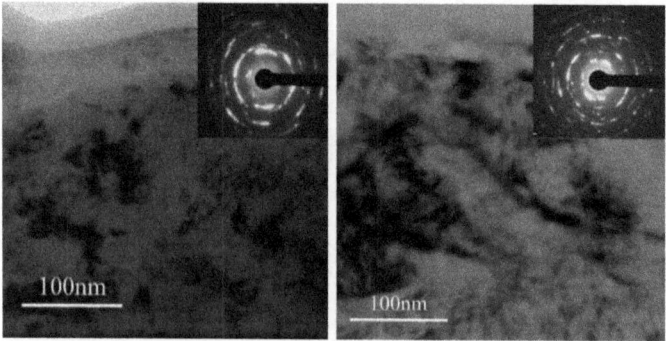

Fig. 5.6 Surface nanocrystals formed by 3 laser shocks

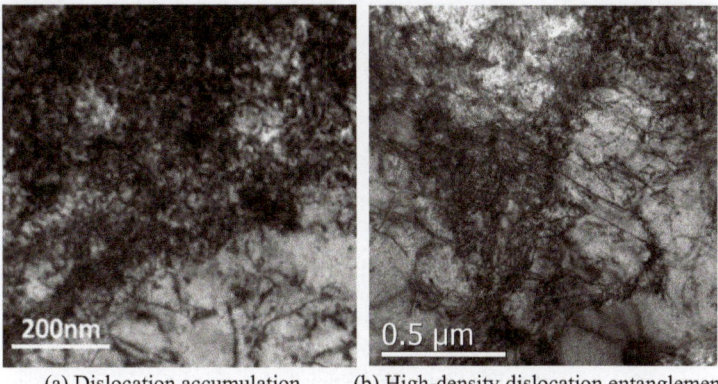

(a) Dislocation accumulation   (b) High-density dislocation entanglement
                                 and dislocation cells

Fig. 5.7 High-density dislocation structure formed by LSP

5.2.3 Compressive Residual Stress Induced by LSP and Its Thermal Relaxation

Among the process parameters, power density and impact times have a great influence on the residual stress field. The GH4133B Ni-based superalloy is treated by LSP with the technological parameters of 5.8, 7.6 GW/cm^2 power density, and 1–3 times of impact. The test results of surface residual stress are shown in Fig. 5.8.

For a GH4133B Ni-based superalloy, a large residual compressive stress is formed on the surface of the material after LSP, where the process parameters have an important influence on the residual compressive stress on the surface. Under the condition of a power density of 5.8 GW/cm^2, the residual compressive stresses formed by 1/2/3 time(s) of impact on the surface of the material are −350 MPa/−428 MPa/−475 MPa,

Fig. 5.8 Surface residual stress under different power densities and impact times

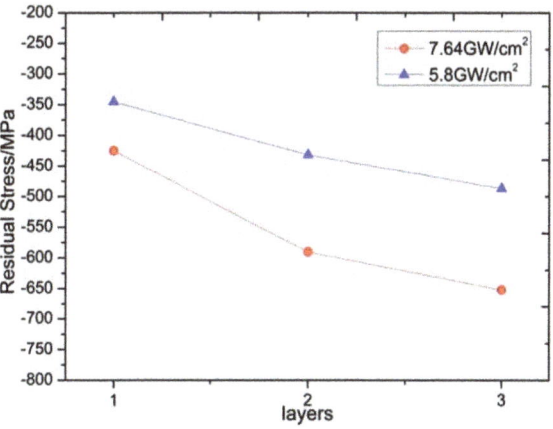

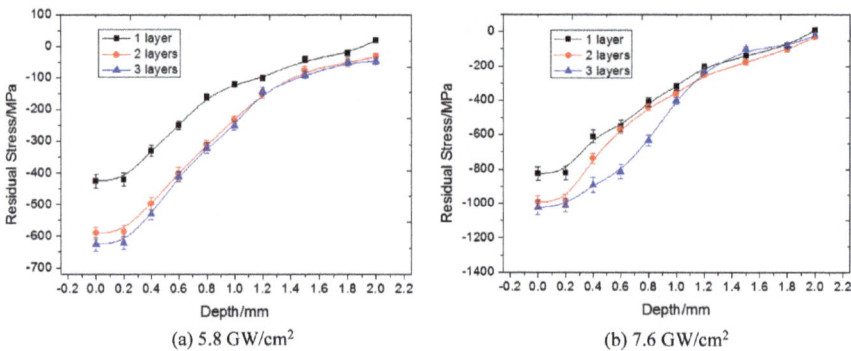

Fig. 5.9 Distribution curve of residual compressive stress along the section under different degrees of power and impact times

respectively. With the increase in the number of impact times, the residual compressive stress on the surface also increases, but the amplitude of increase decreases, which indicates that the plastic deformation induced by multiple LSPtends to become saturated on the surface. When the power density is increased to 7.64 GW/cm^2, the material is treated by 1/2/3 time(s) of impact, the measured surface residual compressive stresses are −427 MPa/−590 MPa/−646 MPa, respectively. It can be seen that with the increase in the power density, the surface residual compressive stress increases, which indicates that the increase in laser energy increases the shock wave pressure and is beneficial to inducing greater plastic deformation.

LSP can not only form a large amount of residual compressive stress on the surface of materials, but it can also change the distribution characteristics of the field of residual stress within a certain depth. The distribution of residual compressive stress along the direction of the depth with power densities of 5.8 and 7.64 GW/cm^2 and 1/2/3 time(s) of impact was tested and is shown in Fig. 5.9.

It can be seen from Fig. 5.9a that when the power density is 5.8 GW/cm^2, the residual compressive stress is formed within a certain depth in the Ni-based super-alloy GH4133B by LSP, and the affected depth exceeds 1 mm. The residual compressive stress test under different impact times shows that the more the impact times, the deeper the affected depth. After 3 times of LSP, the depth of affected layer is 2 mm. At the same depth, the more the number of impacts, the greater the residual compressive stress, but after 1.5 mm, the number of impact times has little effect on the residual compressive stress. Test the residual compressive stress along the direction of the depth with a power density of 7.64 GW/cm^2 and different impact times, as shown in Fig. 5.9b. After increasing the power density, the depth of residual compressive stress affected layer can reach 1.5 mm, because increasing the power density can increase the pressure of the shock wave, thus increasing the depth of affected layer induced by LSP. On the basis of a higher power density, increasing the number of impacts will also increase the amplitude of residual compressive stress and the depth of affected layer, but the improvement effect is not obvious under the condition of a lower power density. Comparing Fig. 5.9a, b, it can be seen that

Fig. 5.10 The relaxation of
surface residual compressive
stress of samples treated by
LSP with different power
densities under different
temperature

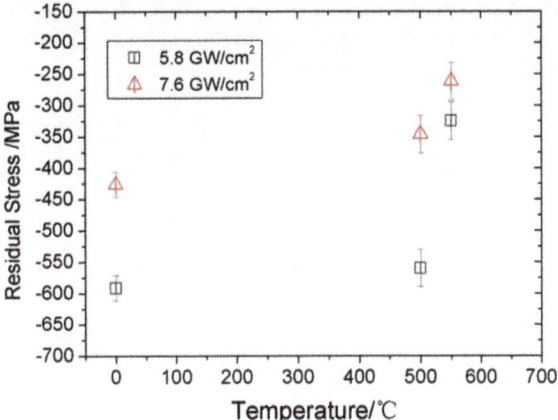

the distribution of residual compressive stress curves under different impact times is similar at the depth of more than 1 mm when the power density is 7.64 GW/cm^2, which indicates that the impact times have little influence on residual compressive stress of GH4133B Ni-based superalloys beyond a certain depth under higher power density.

The main reasons for the occurrence of high-temperature relaxation of residual stress of LSP are as follows: from a macroscopic view, the residual stress is the mutual restraint between the plastic deformation part and the non-plastic deformation part; from a microscopic point of view, residual stress is a manifestation of lattice distortion in metal [30–32]. The working temperature of the GH4133B Ni-based superalloy turbine blade of a certain aero-engine is less than 500 °C. In order to study the thermal relaxation law of Ni-based superalloy after LSP, a heat preservationexperiment was designed. The residual stress of the samples, which were impacted twice at power densities of 5.8 GW/cm^2 and 7.6 GW/cm^2 and kept at 500 °C/1 h and 550 °C/1 h, were tested respectively, as shown in Fig. 5.10.

It can be seen from Fig. 5.10 that the residual compressive stress generated by LSP is partially released under the action of heat. After heat preservation of 500 °C/1 h, the relaxation of the residual compressive stress brought by LSP with a power density of 5.8 GW/cm^2 and 7.6 GW/cm^2 is very small, where only 20 and 7% of the residual compressive stress are released, respectively. However, after heat preservation of 550 °C/1 h, 45 and 60% of residual compressive stress brought by LSP with the two power densities are released, and the higher the temperature, the greater the relaxation of residual compressive stress. We tested the residual compressive stress of the cross section of the samples impacted twice with a power density of 7.6 GW/cm^2 and kept at 550 °C/1 h, as shown in Fig. 5.11.

Under the influence of thermal action, the affected depth of residual compressive stress decreases. The depth decreases from 2 mm before heat preservation to about 0.8 mm. At the same depth, the amplitude of residual compressive stress is also lower than that before heat preservation, but there is still a large residual compressive stress

Fig. 5.11 Distribution of
residual compressive stress
along the depth direction of a
GH4133B blade after heat
preservation

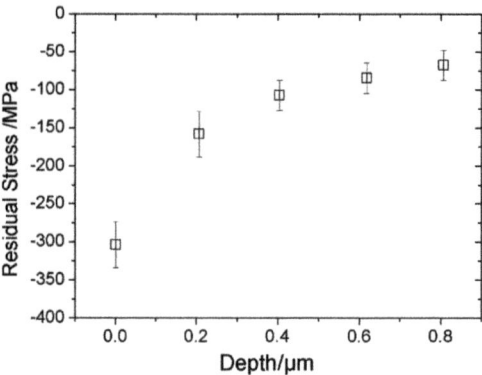

within a certain depth range. The residual compressive stress after thermal relaxation
can still improve the fatigue properties of metals.

5.2.4 Nanohardness and Its Thermal Stability

Nanohardness was used to evaluate the mechanical property of the GH4133B superalloy treated by LSP. The calculation of nanohardness follows the Oliver and Pharr
models [33, 34], with hardness (H) and elastic modulus (E_r) given by the following
formulae, respectively:

$$H = \frac{p_{\max}}{A} \tag{5.1}$$

$$Er = \frac{\sqrt{\pi}}{2\beta} \frac{S}{\sqrt{A}} \tag{5.2}$$

where $p_{\max}$ represents the maximum load, and A represents the contact area. S
represents the stiffness, which is obtained from the slope of the unloading curve, and is
unrelated to the measurement of depth in the indentation process. The cross-sectional
surface of the Ni-based superalloy samples has a certain roughness, which affects the
accuracy of the experimental data. Joslin and Oliver [35] studied the influence of the
surface roughness on nanohardness, and they found that a higher surface roughness
resulted in a larger error between testing and contacting depths of the nanoindenter.
The contact area was closely related to the contact depth, and would cause a larger
dispersion of hardness and elastic modulus results. Therefore, in practical engineering
problems and for some materials with greater surface roughness, Joslin and Oliver
proposed using the H/E_r^2 method to evaluate the mechanical properties of materials.
This will prevent inaccuracy caused by the roughness, and reduce the dispersion of
data. H/E_r^2 represents the capacity of materials to resist plastic deformation.

$$\frac{H}{E_r^2} = \left[\frac{4}{\pi}\right]\left[\frac{P_{max}}{S^2}\right] \qquad (5.3)$$

In this case, H/E_r^2 was also used to correct the elastic modulus of the GH4133B Ni-based superalloy for evaluating the mechanical properties.

Figure 5.12a shows the surface nanohardness of the GH4133B alloy by LSP with different impact times. The nanohardness of the GH4133B alloy without LSP treatment is about 5.6 GPa. After one LSP impact, the surface nanohardness improves by 17.8%, being 7.4 GPa. After three impacts, the surface nanohardness increases to a saturate value, reaching 7.7 GPa. By further increasing the number of impacts, such a value remained unchanged. After 600 °C/10 h thermal treatment, a slight decrease in the surface nanohardness was observed for the three-impact sample. Figure 5.12b, d show variable-depth nanohardness values of the samples with the above treatments. As indicated in Fig. 5.12b, the thickness of affected layer induced by one LSP impact is 100 um, which is consistent with the TEM observation (Fig. 5.12). Within the affected layer, the nanohardness decreases rapidly with a depth of >20 μm. After three LSP impacts, the thickness of the affected layer increases to 600 μm, about 6 times of what it is with one impact (Fig. 5.12c). The nanohardness measured at the 10 μm depth from the surface was 7.7 GPa, and a depth of 100–300 μm shows a value of about 7 Gpa.

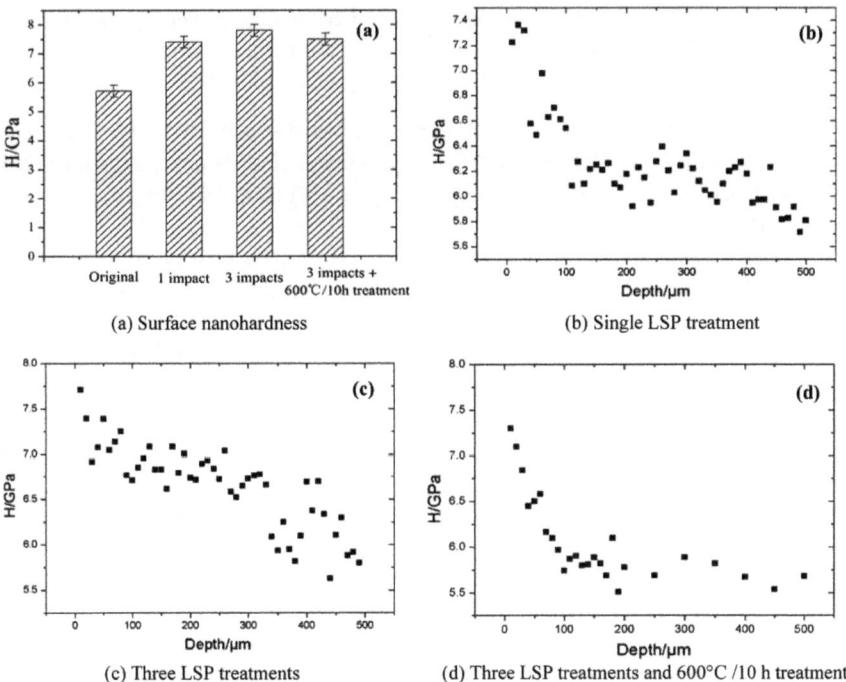

(a) Surface nanohardness

(b) Single LSP treatment

(c) Three LSP treatments

(d) Three LSP treatments and 600°C /10 h treatment

Fig. 5.12 Nanohardness profiles of the GH4133B Ni-based superalloy with different treatments

The sample was polished after receiving three LSP impacts, with a heat preservation of 600 °C/10 h insulation and then furnace cooling. The nanohardness testing after thermal insulation was conducted, as shown in Fig. 5.12d. Like compressive residual stress, thermal activity has a significant influence on the nanohardness. The nanohardness at the 100–600 μm depth from the surface dramatically decreases, which is close to that of the inner part of the sample. However, the surface nanohardness slightly decreases from 7.7 to 7.3 GPa, and the thickness of the hardened layer is about 100 μm [35]. The mechanical properties of Ni-based superalloys depend strongly on their microstructures.

It is well known that increasing dislocation density can effectively improve the hardness of a material [36, 37]. Also, a change in crystal structure leads to an increase in hardness value. An improved model is suggested for the increase in nanohardness by the LSP treatment, as shown in Eq. (5.4):

$$\sigma = \sigma_0 + \sigma_{ds} + \sigma_{RS} + \sigma_{ss} + \ldots \tag{5.4}$$

where σ_0 is substrate nanohardness, σ_{HP} is the nanohardness obtained by the Hall–Petch grain refinement relation, σ_{ds} represents the nanohardness obtained by dislocation strengthening, σ_{RS} represents the nanohardness improved by strain strengthening, and σ_{ss} represents the nanohardness improved by the strengthening phase. For a GH4133B Ni-based superalloy, the nanohardness of its substrate remains unchanged, and the nanohardness increased by LSP is determined by the last four items. The mechanism of improving nanohardness of high-temperature alloys by LSP and the influence of heat treatment are analyzed as follows:

(1)   The strengthening phase of a GH4133B Ni-based superalloy is the γ′ phase. It can be seen from XRD and EBSD analysis that LSP does not change the phase distribution. Therefore, the strengthening effect brought by the strengthening phase can be ruled out;

(2)   Since the selected point for nanohardness is from 10 μm away from the impact surface, while the thickness of the surface nanostructure formed by LSP is only 0.5–1 μm, so, the factor of nanohardness improvement caused by grain refinement obtained by the Hall–Petch relation can also be ruled out;

(3)   The improvement of nanohardness of the Ni-based superalloy by LSP is mainly brought by dislocation strengthening and strain strengthening. Within 100 μm from the impact surface, σ_{ds} and σ_{RS} work together to improve the nanohardness, but beyond 100 μm, σ_{RS} mainly works;

(4)   After heat treatment, the residual compressive stress mostly relaxes, and the strain strengthening factor σ_{RS} caused by the residual compressive stress decreases. Therefore, after heat treatment at 550 °C/10 h, the affected depth of nanohardness decreases to 100 μm, and the rate of the decreasing becomes higher;

(5)   The reason why nanohardness can keep a certain degree of thermal stability within 100 μm from the impact surface is that dislocation strengthening and residual stress work together after thermal action, but dislocation strengthening is the main reason.

5.3 High Temperature High-and-Low Cycle Combined Fatigue Performance of GH4133B Ni-Based Superalloy at 538 °C

5.3.1 High Temperature High-and-Low Cycle Combined Fatigue Performance

It is clear that aero engines undergo a series of stages during flying, e.g., taking-off, climbing, cruising, descending, and landing, which result in low-cycle fatigue damages to the turbine blades owing to varying stress–strain loading. Also, high-cycle fatigue damages at high temperature often happen in blades due to vibration excitation of high-temperature airflow. Therefore, high temperature high-and-low cycle combined fatigue tests were performed for the Ni-based turbine blades in the present work, which is close to the working environment of the actual aeroengine.

During the test, the designed load spectrum was capable of simulating a practical work environment of the assessed positions in the turbine blades, including a high temperature environment and low and high cycle loads. As is shown in Fig. 5.13a, N_L and N_H are low- and high-cycle fatigue numbers, σ_L and σ_H are low- and high-cycle stresses, and f_L and f_H are low- and high-cycle frequencies, respectively. Fatigue testing areas are shown in Fig. 5.13b. Low-cycle loadings are applied on the blade fixture via a hydraulic actuator, while high-cycle ones are imposed by an electromagnetic vibration exciter. Thermal loadings are introduced through an electromagnetic induction coil around the tenon R transfer area with temperatures set to 530 °C.

High-and-low cycle combined fatigue experiments at high temperatures were performed, where six turbine blades were selected for each processing status. Considering real working conditions of aero engineering turbine blades, the peak loading of low cycle fatigue was fixed to be 46.2 kN, and high cycle vibration stress was 337 MPa, and temperature was selected at 530 °C [11]. In this case, fatigue lives of the turbine blades before and after LSP treatment are shown in Fig. 5.14. Interestingly,

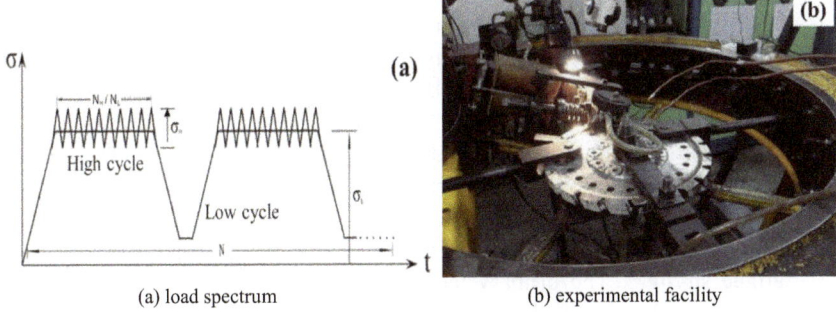

(a) load spectrum   (b) experimental facility

Fig. 5.13 Method of high temperature high-and-low cycle combined fatigue

Fig. 5.14 Fatigue lives of GH4133B alloy turbine blades before and after LSP treatment

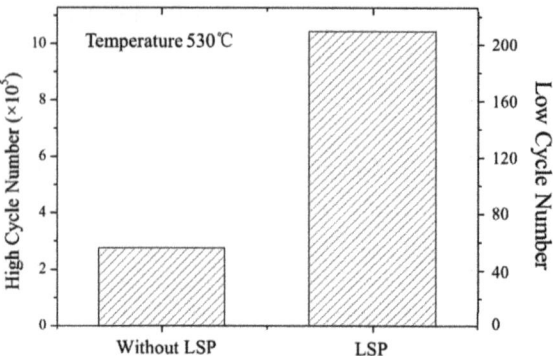

the GH4133B alloy turbine blades with LSP treatment show an excellent fatigue-resistant capacity. Their high-cycle fatigue lives increase by 3.8 times as compared to that without LSP treatments (from 275,073 to 1,043,216) [36].

LSP can effectively improve the fatigue performance of nickel-base alloy parts. By analyzing its strengthening mechanism, the main reasons are as follows:

(1) Surface nanocrystals can effectively improve the nucleation life of high-cycle fatigue cracks in Ni-based superalloy. For the high cycle fatigue process, the crack nucleation process takes about 90% of the fatigue life of materials, so the fatigue life of materials mainly depends on the crack nucleation time. The process of fatigue crack nucleation by the dislocation pile-up model could described as following:

$$n_c = Ka^{-1/2} \tag{5.5}$$

where n_c the crack nucleation life, a is the grain radius, $K = \frac{\sigma\sqrt{2h}}{3(\Delta\tau-2b)}$, h is the distance between dislocation pile-up planes, $\Delta\tau$ is the maximum and minimum stress difference, σ is the front stress of the dislocation pile-up, b is the Burgers vector.

It can be seen from Formula (5.5) that K can be considered as a constant when the materials are the same and the applied load is determined, and the number of crack nucleation cycles increases with a decrease in grain size. Previous studies have shown that there is a definite correlation between high cycle fatigue strength and slip length. The usual mechanism of fatigue crack nucleation is closely related to the interaction between the slip band and the specimen surface or other boundaries (such as the grain boundary or the phase boundary). A decrease in all microstructure sizes will cause a decrease in the length of the slip, thus increasing crack nucleation resistance and high cycle fatigue strength. The primary factor affecting the length of the slip of the structure is the size of the structure. Therefore, the reduction in the size of the structure can directly reduce the length of the slip and improve the fatigue strength.

For the surface nanolayer induced by LSP, the size of the nano-grain is smaller than the characteristic size of dislocation spacing. Under the same amount of deformation, the more grains in a certain volume of metal, more grains the deformation is dispersed in, so that the deformation distribution is more uniform, thus the stress concentration that may lead to local cracking is less, so that the material can bear a larger amount of deformation and show better plasticity before fracture. In a high temperature environment, the surface gradient structure induced by LSP can also effectively improve the high temperature oxidation performance and prolong the process of crack nucleation.

(2) High-density dislocations can improve the work hardening effect of the Ni-based superalloy. The surface nanocrystallization is the result of intense plastic deformation [6, 36, 37], which will cause high-density dislocation, grain boundary and other crystal defects in the deformed structure. These defects can hinder the movement of dislocations in metal crystals and make it difficult for the plastic flow of metal materials to occur, thus improving the strength of metal.

The mechanical properties of materials are closely related to the geometrical shape of dislocations, the interaction among dislocations, the interaction between dislocations and solute atoms, and the movement of the dislocations. The plastic deformation of Ni-based superalloy is mainly coordinated by dislocation movement. When dislocation moves on the slip surface, it is blocked by grain boundaries and piles up. Only when the stress concentration caused by a dislocation pile-up group increases to a certain extent, can the adjacent grains be forced to slip correspondingly, causing the macroscopic effect of plastic deformation, and correspondingly increasing the yield strength of the Ni-based superalloy. For metal crystals, the increase in strengthening caused by dislocations is roughly proportional to the square root of dislocation density in the crystal, that is,

$$\Delta\sigma_d \propto \rho^{1/2} \tag{5.6}$$

where, $\Delta\sigma_d$ represents the strengthening increment, ρ represents the dislocation density in the metal. A large number of research results show that work hardening is mainly to improve the resistance to fatigue crack formation, while residual compressive stress is mainly to increase the resistance to fatigue crack propagation. If softening or residual tensile stress occurs in plastic deformation, the effect is just the opposite, especially if the tensile stress concentration is caused by uneven deformation in the micro region during deformation, whose harm cannot be ignored.

(3) Nanocrystals weaken the influence of stress concentration. The fatigue crack initiation is closely related to the surface state of materials, and it is easy to produce surface unevenness and micro-notch during processing. The stress concentration caused by a change in roughness overlaps with the concentration of stress at the notch, which intensifies the concentration of stress at the root of the notch and promotes the crack generation. Therefore, the increase in

surface roughness reduces the fatigue limit of materials. And the concentration of stress is higher at the slip band, grain boundary, inclusion of metal surface or the interface between the second phase and the substrate. Therefore, fatigue cracks generally originate from the surface of a metal part. LSP basically does not change the roughness of metal materials, and after surface nanocrystallization, the slip band, grain boundary and second-phase interface on the surface are relatively reduced, which is conducive to inhibiting the initiation of cracks. Grain refinement is equivalent to reducing the average slip distance, and reducing the concentration of stress caused by dislocation pile-up at grain boundaries.

5.3.2 Observation and Analysis of Fatigue Fracture of Ni-Based Superalloy Turbine Blades

LSP-induced gradient structure can effectively improve the fatigue life of turbine blades. In this section, the fatigue source areas strengthened by LSP are observed by TEM, and the stability of surface nanocrystals under the combined action of high temperature oxidation and fatigue loading is investigated. Cross-section TEM samples are extracted by FIB from the fatigue source area of the R slot fracture of turbine blades, as shown in Fig. 5.15.

Observe the surface source area, as shown in Fig. 5.15. It can be seen that the surface nanocrystals induced by LSP still exist after high temperature and high-cycle fatigue loading, and the grain size has no obvious change compared with that before fatigue loading. The surface nanocrystals obtained by mechanical grinding will grow under mechanical loading, and with the increase in the strain rate and strain capacity, the growth will become obvious. This shows that the surface nanocrystals induced by LSP have better stability under fatigue loading than those formed by mechanical grinding (Fig. 5.16).

Fig. 5.15 A TEM observation of a cross section of a fatigue source area of a strengthened turbine blade

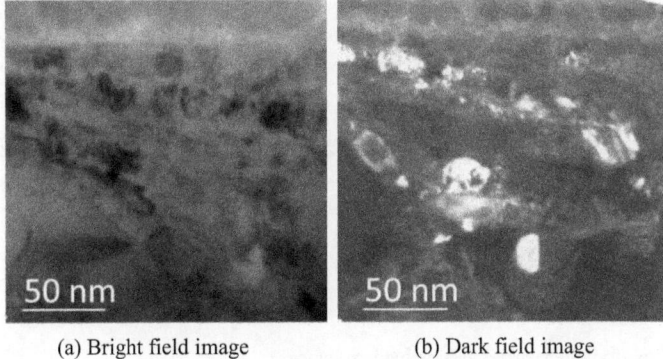

       (a) Bright field image   (b) Dark field image

Fig. 5.16   TEM observation on the surface of a turbine blade after high temperature fatigue

When observed at a distance of 2 μm from the impact surface, high-density dislocations can still be found, as shown in Fig. 5.17. In the high-density dislocation region, due to the weak interface effect, the crack propagation is blocked by the interface, and the crack tip is prone to passivation, which reduces the fatigue crack propagation rate.

By analyzing the fracture surface after LSP, it was found that all specimens of turbine blades made of the GH4133B Ni-based superalloy were fractured from the R place of the first tenon, with obvious fatigue failure characteristics, and the fatigue area accounts for 30–40% of the total fracture surface area, as shown in Fig. 5.18.

The turbine blades before and after LSP were selected for fracture observation, and the influence of LSP on fatigue process was analyzed. The blades that were not strengthened by LSP were observed. The fatigue fracture of a GH4133B turbine blade shows clear cleavage-river-pattern and high cycle and low cycle fatigue bands perpendicular to the cleavage-river-pattern, as shown in Fig. 5.19. The mechanism of the formation of the "river-pattern" is that when the turbine blade is loaded, the

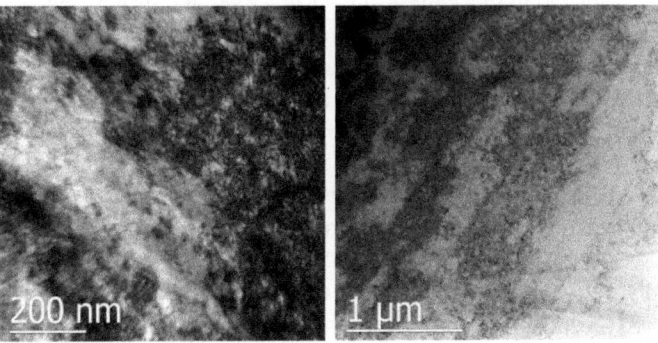

Fig. 5.17   High-density dislocations on the secondary surface of strengthened turbine blades after high temperature fatigue

Fig. 5.18 Situation after partial turbine blade fracture

    (a) Secondary electron imaging   (b) Backscattered electron image

Fig. 5.19 Macroscopic fracture

shear stress on all slip systems increases, but the slip starts only on the system that reaches the critical decomposition shear stress of this material. With the propagation of cracks, these horizontal planes combine to form a crack, which advances on fewer parallel planes, resulting in a pattern similar to a river and its branches.

The time of high cycle fatigue crack initiation is about 90% of the fatigue life of the material, so the fatigue life of materials mainly depends on the time of crack initiation, that is, the process of crack nucleation. The fatigue source area generally includes fatigue crack initiation and short crack propagation. There is only one fatigue source area on the turbine blade, which is located on the surface of the sample, 0.5 mm away from the exhaust edge, indicating that the internal defects of the turbine blade are few and the material is uniform, as shown in Fig. 5.20a. By analyzing the formation of cracks, it is considered that the maximum stress of a blade under high cycle fatigue vibration loading lies on the surface of the blade, and the free surface grains of members are weakly restrained by other grains, which makes it easy to produce plastic deformation, especially on the free surface of the slip of the grains with the highest orientation, which causes crack initiation. The micromorphology of the fatigue source area and the area of early crack propagation is

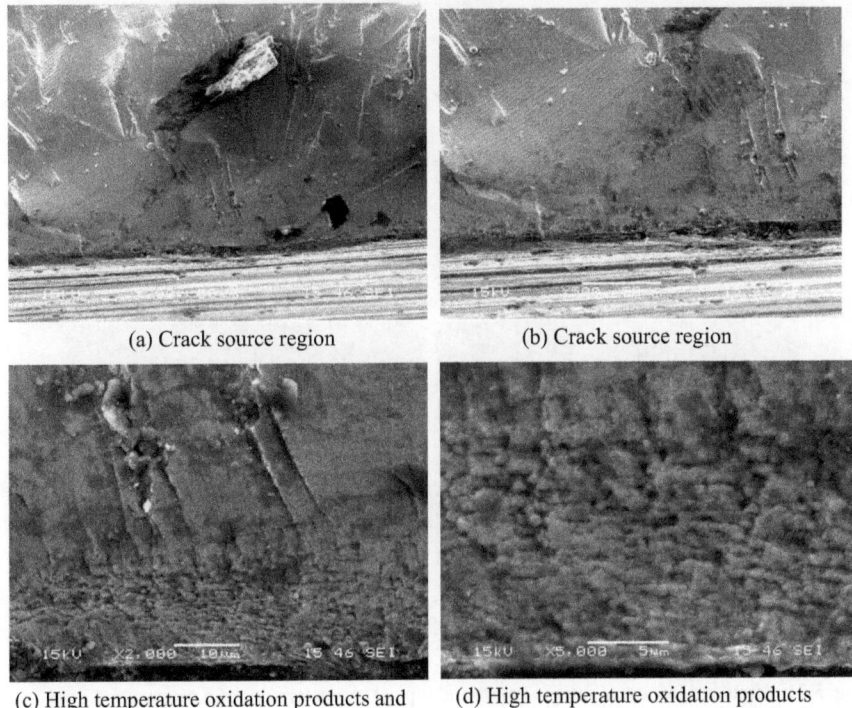

(a) Crack source region

(b) Crack source region

(c) High temperature oxidation products and cleavage characteristics

(d) High temperature oxidation products

Fig. 5.20 Features and typical characteristics of crack source area and early fatigue propagation area

complex, showing the slip line, cleavage morphology, and an early fatigue band, and usually also transgranular and intergranular mixed fracture characteristics, as shown in Fig. 5.20b.

Turbine blades are subjected to high-temperature and high-cycle combined fatigue. Under the influence of high temperatures, the fatigue source area presents a large number of high-temperature oxidation characteristics. In Fig. 5.20c, d, some randomly distributed oxides can be seen in the source area, and these oxidation products can accelerate crack initiation and nucleation and reduce fatigue life.

Figures 5.21a, b are the typical features of crack propagation areas, showing a large number of cleavage features, and obvious cleavage planes and river-patterns can be observed. The crack propagation happens in a way that slip planes are separated due to periodic slips. In the transition area between the region of fatigue source and the steady-state propagation area, there are early fatigue bands. Under the action of high temperatures, the high cycle and low cycle fatigue bands are oxidized, and their contours are clearer. It is easy to measure the spacing of low cycle fatigue bands under the electron microscope. In Fig. 5.21c, the low cycle fatigue band spacing at 200 μm from the source region is about 100 μm. Along the crack propagation

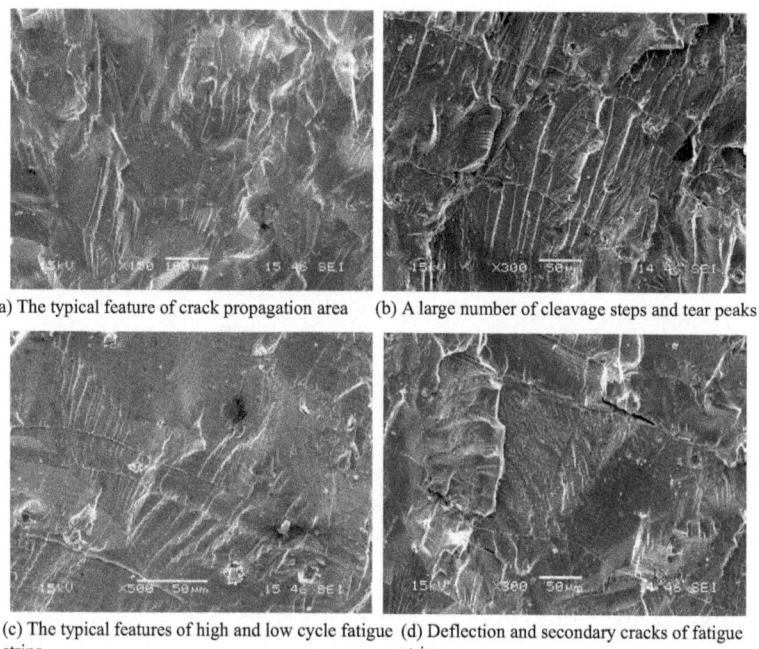

(a) The typical feature of crack propagation area   (b) A large number of cleavage steps and tear peaks

(c) The typical features of high and low cycle fatigue (d) Deflection and secondary cracks of fatigue
    strips   strips

Fig. 5.21 High and low cycle fatigue bands and typical characteristics of a propagation area

direction, the fatigue strip spacing becomes wider. This is because with the increase in the stress intensity factor ΔK, the crack growth rate increases.

In the process of crack propagation, it encounters the second-phase particles and grain boundaries, and deflects at the interface, as shown in Fig. 5.21d. Second-phase particles hinder crack propagation to some extent. When fatigue cracks pass through larger-sized second-phase particles, fatigue bands are also formed on the surface of the second-phase particles, and the fatigue bands become dense. This is because the plastic zone needs to be re-formed at the crack tip before the new crack can continue to propagate, and the existence of two-phase particles slows down the rate of crack propagation. Secondary cracks are also observed in the crack propagation area, which is due to the concentration of stress at some impurities, second-phase particles and grain boundaries, which tend to cause the tearing of the GH4133B Ni-based superalloy along the crack surface, thus forming secondary cracks.

Generally, the fast fracture area of fatigue fracture presents the appearance of mixed fracture, and the grain boundary strength of the GH4133B Ni-based superalloy is lower than the intragranular strength, so the fast fracture area presents the mixed fracture mode of intergranular and dimple, and gradually changes from a ductile fracture to a quasi-brittle fracture, as shown in Fig. 5.22a. The fracture surface presents an intergranular dimple fracture mode, as shown in Fig. 5.22b.

The fracture of a turbine blade subjected to LSP is observed and compared with the fatigue macroscopic fracture of the untreated blade (Fig. 5.23). After the turbine

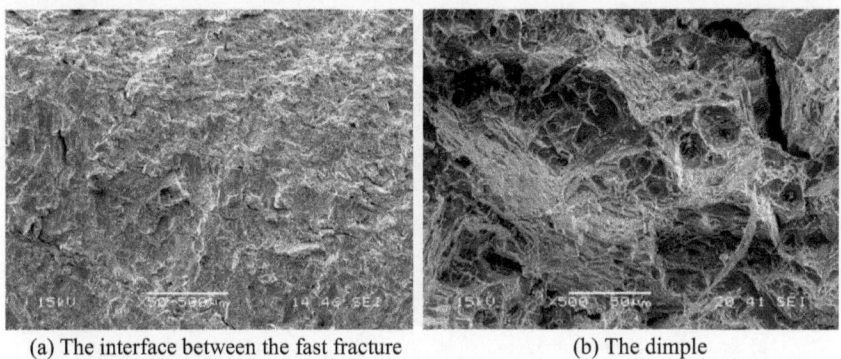

(a) The interface between the fast fracture   (b) The dimple
area and the propagation area

Fig. 5.22   Crack features and typical characteristics of the fast fracture area

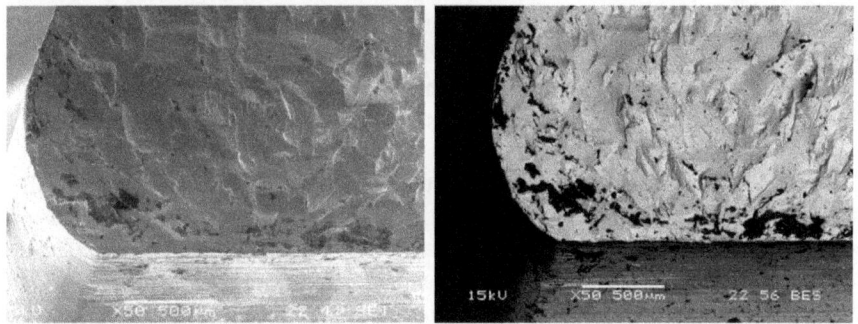

(a) Macro-feature secondary electron imaging (b) Macro-feature backscattered electron image

Fig. 5.23   Macroscopic feature of the fracture surface of a turbine blade after LSP

blade is subjected to LSP, the fracture feature of the specimen is still dominated by a cleavage river-pattern and fatigue band, and the mechanism of crack initiation and propagation has not changed obviously compared with that of the untreated blade, but the whole fracture distribution is relatively flat compared with that of the untreated one, with relatively few cleavage steps and a larger crack propagation area, as shown in Fig. 5.23.

Figure 5.24 shows the feature of the crack source area. It can be observed that there is only one region of the crack source which is still located at the surface 0.5 mm away from the exhaust edge, Compared with the sample without LSP treatment, the crack source area is relatively flat, and except for the early fatigue bands and inter-granular fracture characteristics in the crack source area, only a few high-temperature oxidation products are observed, which indicates that LSP could effectively improve the high-temperature oxidation resistance of materials.

When studying the high temperature oxidation resistance of high-temperature alloys, the key lies in whether a dense protective oxide film can be quickly formed

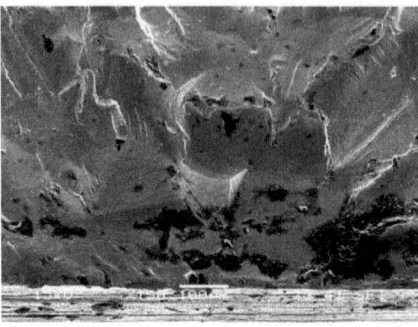 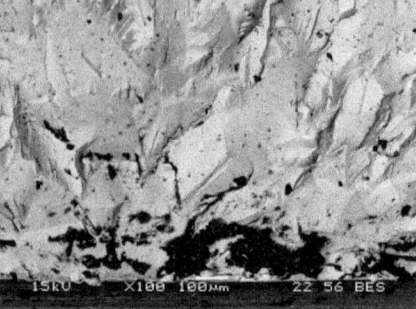

(a) Secondary electron imaging   (b) Backscattered electron image

Fig. 5.24   Features of the crack source area

on the surface in the initial stage of oxidation. There are many pieces of research on improving the high temperature oxidation resistance of Ni-based superalloy by surface nanocrystallization, which can be summarized as follows: (1) After surface nanocrystallization of the Ni-based superalloy, crystal defects such as grain boundaries and dislocations on the surface of samples increase, and they have a higher surface free energy. In the initial stage of oxidation, it can promote the nucleation of oxide and provide more nucleation sites, resulting in rapid oxidation; (2) After surface nanocrystallization, there is a great number of channels for element diffusion, which promotes the rapid diffusion of alloy elements such as Cr from the substrate to substrate/oxide interface in the oxidation process, thus improving the content and density of Cr in the surface oxide film and improving oxidation resistance; (3) Oxides nucleate and grow first at crystal defects, and the refinement of surface grains will also lead to the refinement of the generated oxide grains, which will form pinning at crystal grain boundaries, thus improving the adhesion and toughness of the oxide film to the substrate, and further improving the resistance to high temperature oxidation.

Figure 5.25 shows the feature of the crack propagation area. It can be observed that the spacing distribution of low cycle fatigue bands after LSP is narrow, and the low cycle fatigue band spacing at 370 μm away from the source area is 12 μm (Fig. 5.25a), which is obviously smaller than the sample before LSP (Fig. 5.21c), indicating that LSP inhibits crack propagation. In addition, it can be seen that the crack propagation area is relatively flat, with few cleavage planes and secondary cracks, as shown in Fig. 5.25b. Different grain orientations also have a certain influence on crack propagation. When the crack propagates to a grain boundary and the fatigue band deflects, as shown in Fig. 5.25c, there is a twinning structure in the GH4133B Ni-based superalloy, and the regular distribution of fatigue bands can be seen in the substrate on both sides of the twin band and inside, as shown in Fig. 5.25d. The influence of second-phase particles on crack propagation is also observed in the fracture surface of the turbine blade strengthened by LSP. As shown in Fig. 5.25b, d, it can be observed that the fatigue strip spacing is obviously narrowed after the crack passes through the second-phase particles.

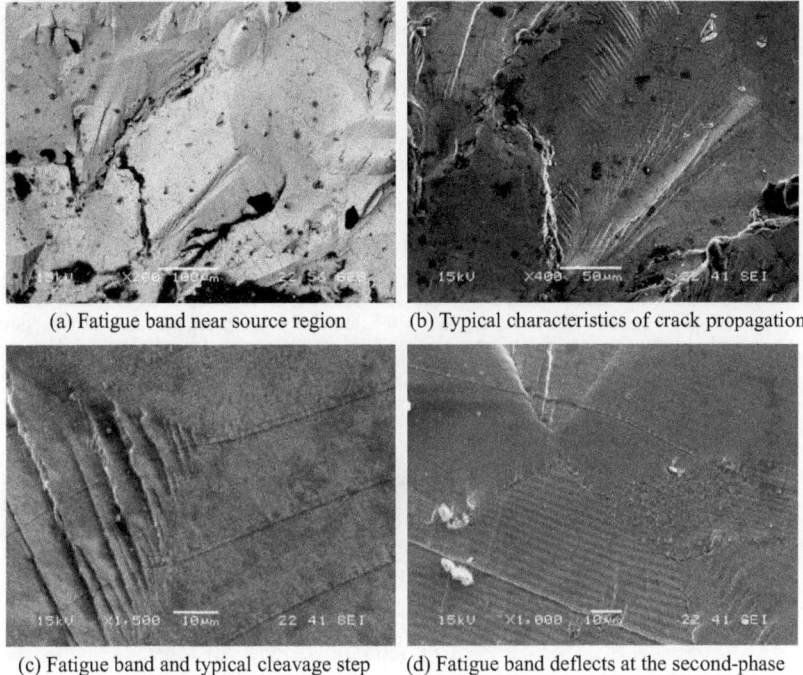

(a) Fatigue band near source region   (b) Typical characteristics of crack propagation

(c) Fatigue band and typical cleavage step   (d) Fatigue band deflects at the second-phase
                                              particles

Fig. 5.25 Typical characteristics of the crack propagation area after LSP

Figure 5.26 shows the feature of the fast fracture area, which is similar to the fracture feature of the untreated blade, where the fracture mode of the fast fracture area gradually changes from ductile fracture to quasi-brittle fracture, as shown in Fig. 5.26a, b. The fracture surface also presents an intergranular dimple fracture mode, and a fracture plane and a secondary crack can be observed at the bottom of the dimple, as shown in Fig. 5.26c. It can be seen from Fig. 5.26d that the crack propagation does not pass through the back of the mortise.

According to the fracture analysis, the fracture of the turbine blade made of GH4133B Ni-based superalloy consists of three parts: fatigue source area, steady-state propagation area, and fast fracture area. The fracture surface presents a clear cleavage river-pattern and high cycle and low cycle fatigue strips perpendicular to the cleavage river-pattern. Under the action of high temperatures, the fatigue strips are oxidized, deflected at the second-phase particles, and the width of fatigue strips is narrowed.

The LSP did not change the initiation and propagation mechanism of high-temperature and high-and-low cycle combined fatigue crack of the GH4133B Ni-based superalloy turbine blade. However, it has the following effects: (1) surface nanocrystals induced by LSP improve the resistance of Ni-based superalloy to high temperature oxidation; (2) the spacing of low cycle fatigue strips after LSP is much

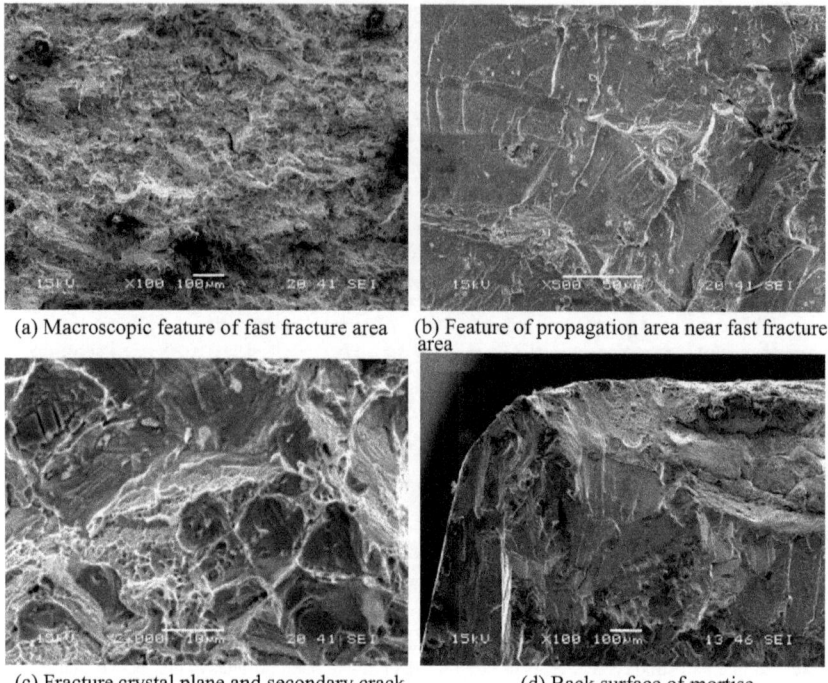

(a) Macroscopic feature of fast fracture area   (b) Feature of propagation area near fast fracture area

(c) Fracture crystal plane and secondary crack   (d) Back surface of mortise

Fig. 5.26 Feature of the fast fracture area

narrower than that before LSP. From the previous research, it is known that the high dense dislocation produced by LSP and the residual compressive stress after thermal relaxation can effectively inhibit crack propagation; (3) after LSP, the crack propagation area of the blade is flat compared with what it was before LSP, and the crack propagation area is larger and cleavage features are relatively fewer.

5.4 Gradient Microstructure Characteristics Induced by LSP in the K417 Ni-Based Superalloy

5.4.1 The K417 Ni-Based Superalloy and Experimental Procedure of LSP

The K417 Ni-based superalloy was chosen as another subject in this work, which also has been widely applied for turbine blades in aero-engines. The coarse dendritic structure with the size from several hundred micrometers to several millimeters can be found in Fig. 5.27a. Figure 5.27b shows the as cast sample structure, which mainly consists of γ solid solution, secondary γ' phase and (γ + γ') eutectic phase.

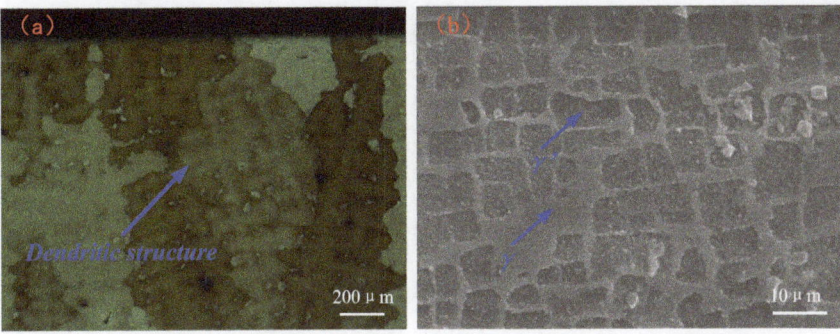

(a) Optical microscope (OM) observation   (b) Scanning electron microscopy (SEM) observation

Fig. 5.27 The microstructure of the K417 Ni-based superalloy

The γ solid solution is the matrix with a continuously-distributed face-centered cubic structure. The large amount of γ' phase distributing among the dendrite are the main strengthening phases. The nominal chemical composition of the K417 Ni-based superalloy is shown in Table 5.3.

The samples of the K417 Ni-based superalloy for LSP treatment were taken from the groove parts and then cut into rectangular shapes in dimensions of 25 × 20 × 4 mm (width × length × thickness), which are schematically shown in Fig. 5.28. Prior to LSP treatment, the surface of the samples was polished with SiC paper with the grit number from 500 to 2400, and then ultrasound cleaning was used to degrease the surface of samples in ethanol. The paths of the laser spots and the LSP region are shown in Fig. 5.28, and the detailed laser parameters are listed in Table 5.4 [38].

5.4.2 Gradient Microstructure Induced by LSP and Its Thermal Stability

TEM images obtained from the top surface layer (about 0.5 μm thick) of the LSPed region are shown in Fig. 5.29. Figure 5.29a shows the dual-phases with matrix phase γ and second phase γ' in the original sample, which is consistent with the original SEM images. The corresponding SAED also shows the dual-phase structure. In Fig. 5.29b, c, the C deposition layer above the sample surface with a good integrity shows that there is no damage formation during the process of the preparation of the sample. It can be found that a surface nanostructure with the grain sizes of about 20–200 nm was produced after LSP treatment, as seen in Fig. 5.29b. The corresponding SAED pattern is dominated by circles, which indicates that the orientations of the nanocrystallines are random and many of these nanocrystallines have high angle grain boundaries. In addition, the nanocrystallines were obviously elongated perpendicular to the peening direction. The dark-field image is presented in Fig. 5.29c, and the corresponding grain size distributions indicate that only a small fraction of the microstructure is made

Table 5.3 Composition of the K417 Ni-based superalloy

Composition	Cr	Co	Mo	Ti	Al	C	V	B	Zr	Mn	Si	Ni
Percentage (wt.%)	8.5–9.5	14–16	2.5–3.5	4.5–5.7	4.8–5.7	0.13–0.22	0.6–0.9	0.012–0.022	0.05–0.09	<0.5	<0.5	Bal.

Fig. 5.28 The samples of K417 for LSP treatment and scanning path

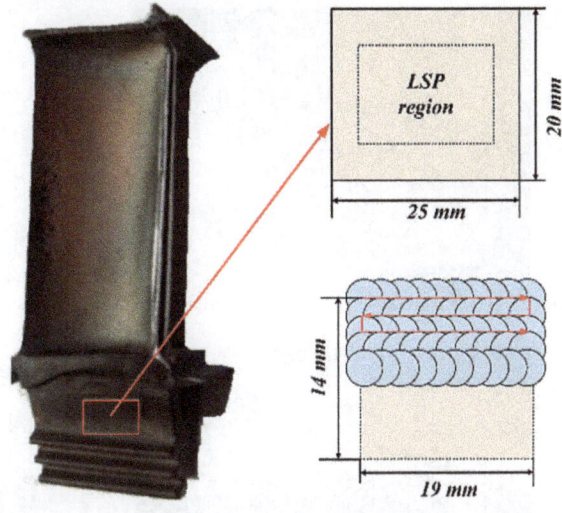

Table 5.4 The detailed laser parameters

Parameters	Values
Laser wavelength (nm)	1064
Pulse energy (J)	10.8
Pulse duration (ns)	20
Spot diameter (mm)	3.4
Repetition rate (Hz)	1
Lapping rate	60%
Impacts time	5

up of grains larger than 100 nm, as shown in Fig. 5.29 (d). It is to note that the coarse dendritic structure disappears and the phase boundaries of γ and γ' cannot be observed in this region. In addition, a high dense dislocation can be observed within the subsurface layer of the K417 Ni-based superalloy with a depth of 1–5 μm in our previous study [10]. In this region, the phase boundaries of γ and γ' can be observed. Due to the existence of the channels of the dislocation motion in the γ phase, higher density dislocation can be found in the γ phase. The characteristics of the gradient microstructure in the deformation layer are a result of the laser-induced shock wave pressure decay along the depth [39–41]. Due to the largest plastic deformation induced by LSP on the surface layer, the nanoscale grain was formed. With an increase in the depth from the surface, the decay of shock pressure is not enough to induce the formation of nanocrystallines [3, 42, 43]. That is why a high dense dislocation structure can be found only in this region (1–5 μm depth).

A comprehensive understanding of the mechanism of surface nanocrystallization induced by LSP has been described in detail in literature [44]. For the K417 Ni-based superalloy, dislocation activities are motivated in the original coarse grains

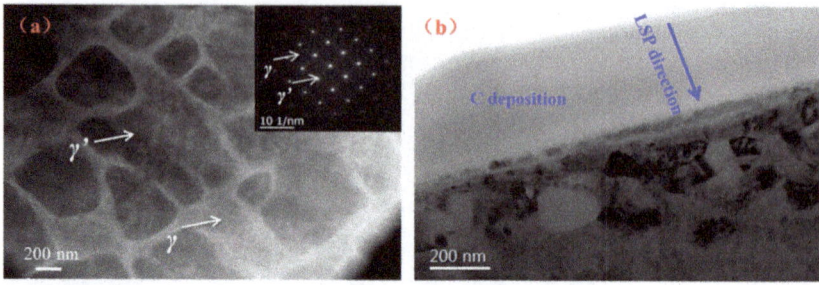

(a) The STEM microstructure feature of the original sample   (b) surface nanostructure on the top surface layer

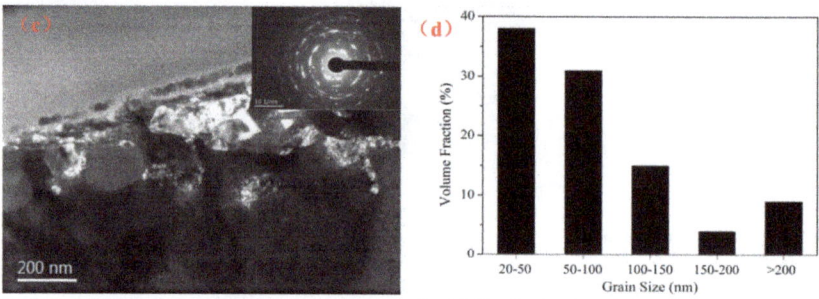

(c) The corresponding TEM dark-field image of image (b)   (d) The distribution of corresponding grain sizes

Fig. 5.29 Characteristics of surface microstructure with and without LSP treatment

during the plastic deformation. According to the homogeneous nucleation theory proposed by Meyers [45], high density dislocations were formed at the shock wave front when the shock wave pressure reaches a critical level. Along with the increase in shock wave pressure, dislocation walls and dislocation tangles were generated and further subdivided the original grain by realizing dislocation cells. Dislocation rearrangement and annihilation will be generated when strains increase up to a certain threshold, which transforms the original coarse grains into the sub-structure in the original grains. Eventually, nanoscale grains with random orientation were formed on the top of the surface under continuous plastic deformation.

Generally, the Ni-based superalloy is used at a particularly high temperature. In a previous study [10], we found that there is a beneficial effect of LSP on prolonging the fatigue strength of K417 Ni-based superalloys from 110 to 230 MPa at a temperature as high as 900 °C. In addition, we found the compressive residual stress in the surface layers had been almost completely relaxed with about 72% at this temperature and the surface nanostructure had kept a good thermal stability. These results suggest that the near-surface nanostructure plays a more significant role in improving the fatigue strength at elevated temperatures. However, the mechanism of thermal stability of surface nanostructure produced by LSP has not been discussed. Thus, the characterization of the microstructure after annealing treatment was adopted to reveal the mechanism of thermal stability in this section. The LSP samples obtained

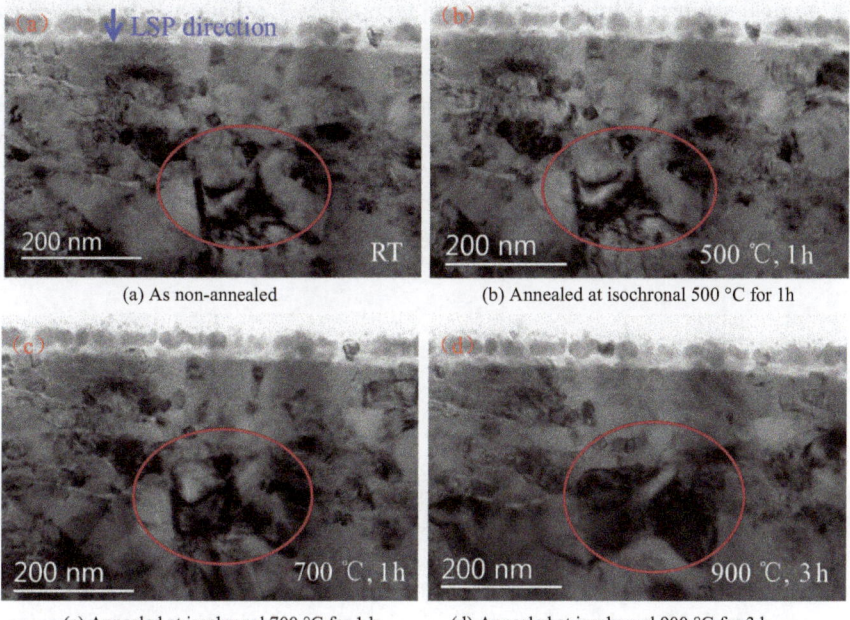

(a) As non-annealed   (b) Annealed at isochronal 500 °C for 1h

(c) Annealed at isochronal 700 °C for 1 h   (d) Annealed at isochronal 900 °C for 3 h

Fig. 5.30 TEM images showing the change in surface nanostructure in the Ni-based superalloy under different annealing conditions

by FIB were annealed by in-situ TEM annealing at 500 °C for the duration time of 1 h, 700 °C for the duration time of 1 h and 900 °C for the duration time of 3 h, respectively.

The evolution of the microstructure during the in-situ TEM annealing experiment is shown in Fig. 5.30, the grains with nanoscale sizes on the surface layer can be clearly found in each picture. These images were taken from the same region of the thin foil, so the real time evolution is apparent. Laser-induced severe plastic deformation results in the significant grain refinement (see in Fig. 5.30a). After LSP treatment, the grain sizes were reduced to about 20–140 nm and the microstructure consisted of both equiaxed and elongated grains perpendicular to the LSP direction, as shown in Fig. 5.30a. When the temperature increases to $0.38 T_m$ of 500 °C (T_m: melting point of the K417 Ni-based superalloy is 1300 °C), there is no change in grain size and grain boundary compared with that in Fig. 5.30 (a), as shown in Fig. 5.30b. The annealing temperature is further increased to $0.54 T_m$ of 700 °C with the duration time of 1 h, and the microstructure in Fig. 5.30c looks similar in terms of grain sizes of the nanocrystallines obtained from the as-deformed condition in Fig. 5.30a. In other words, the surface nanocrystalline remains comparatively stable at 700 °C for 1 h. It indicates that the surface nanostructure induced by LSP has a higher critical growth temperature than the dynamic crystallization temperature, $0.36 T_m$ of 468 °C for original samples. Similar results were reported by Lewandowska et al. [46]. It is notable that the grain boundaries become blurred after 700 °C of 1 h annealing treatment, which reveals that the grain has a tendency to grow at this temperature. Figure 5.30d is taken after 900 °C for 3 h annealing, the obvious changes in grain sizes are observed at this high temperature and the largest grain size increases from about 140 nm to about 220 nm, as seen in the red loop in Fig. 5.30. Moreover, recrystallization occurs in this condition. However, some smaller grains with around 50 nm can still be found in this image (Fig. 5.30d), which indicates that the surface nanostructure has a good thermal stability at 900 °C.

5.4.3 *Compressive Residual Stress Induced by LSP and Its Thermal Relaxation*

The residual stress distributions as a function of depth are presented in Fig. 5.31. In 6 GW/cm^2 treatment, the surface compressive residual stress for a single LSP impact is −583 MPa and 3 LSP impacts is −624 MPa. Meanwhile, the peak compressive residual stress for a single LSP impact is −667 MPa and 3 LSP impacts is −702 MPa. It can be seen that both the surface and peak compressive residual stress are increased by 7 and 5% when the impact time increases from 1 to 3. And the depth of compressive residual stress reaches about 800 μm in 1 impact. As for the sample with a treatment of 3 impacts, the depth reaches about 1000 μm, has apparently increased by nearly 25%. The cumulative impacts have a very superficial effect on the residual stress levels in the superficial layers. So, it can be concluded that the impact times have a more important influence on the residual stress. The residual stress may counteract some or all of the tensile stress, decrease the crack propagation rate, effectively reduce the stress intensity factors, enhance the fatigue crack closure effect, and increase the

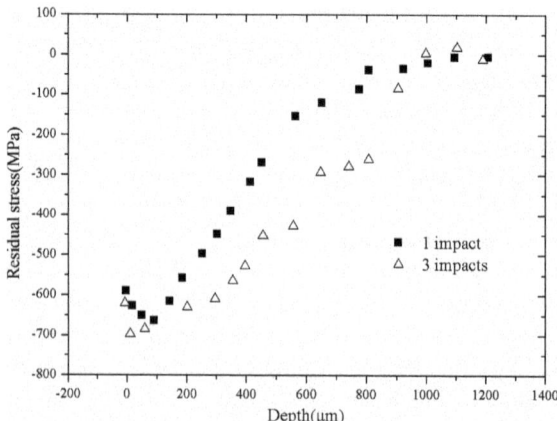

Fig. 5.31 Residual stress profiles of the hardening layer after multiple LSP impacts in 6GW/cm^2

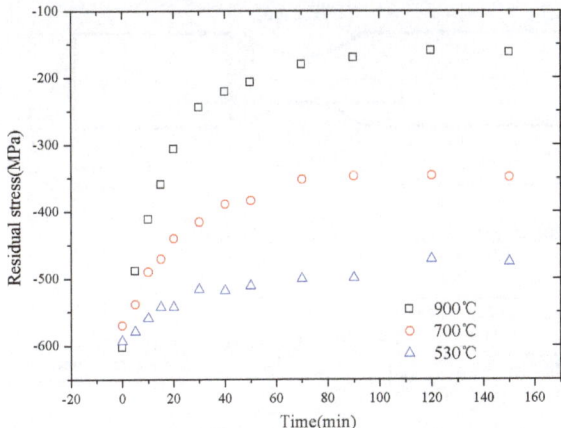

Fig. 5.32 Surface residual compressive stress relaxation curve of the K417 alloy annealed at different temperatures

critical stress of crack propagation, thus improving the fatigue performance of metal materials. In order to assess the residual stress thermal relaxation of LSP, the heat treatment was conducted after LSP.

As shown in Fig. 5.32, the surface compressive residual stress induced by a single LSP impact gradually decreases with the lasting thermal holding time at different temperatures. The temperature has a significant influence on the compressive residual stress. After 530°C/700°C/900°C for 150 min heat treatment, the surface residual compressive stress slightly declined from 595 MPa/568 MPa/582 MPa to 482 MPa/347 MPa/161 MPa. It can be seen that the surface compressive residual stress releases by 19% and 39% when the temperature increases from 530 to 700 °C, and 72% compressive surface residual stress has a relaxation when the temperature increases to 900 °C. After the first 30 min for 900 °C heat treatment, the residual stress releases quickly, with 60% compressive residual stress reaches relaxation, and then the residual stress continued to reduce, but the decreasing trend would be gradually slow, and would still become basically stable.

5.5 High Temperature High Cycle Combined Fatigue Performance at 800 °C

The K417 Ni-based superalloy turbine blade failure forms commonly used for vibration fatigue. The vibration fatigue test was designed, and the shape and size of the test sample is shown in Fig. 5.33. The dashed rectangular part in the specimen is the original crack zone where the LSP treatment is carried out.

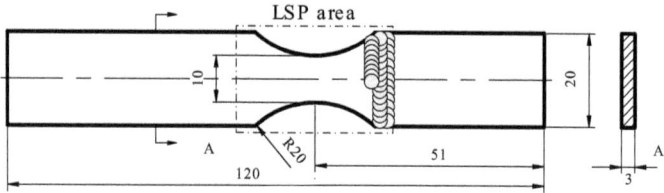

Fig. 5.33 The shape and size of test specimen and the LSP area

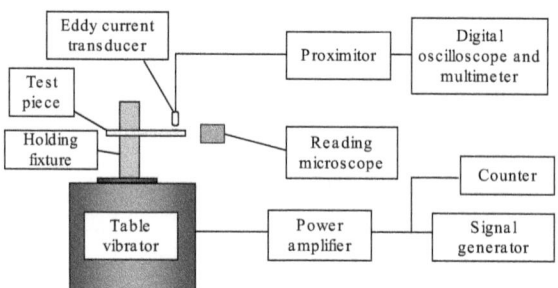

Fig. 5.34 Vibratory experiment system

In order to study the influence of the thermal effect to LSP, we carried out a comparison of the vibration fatigue performance between the "LSP" and "LSP+ thermal insulation" in which an up and down method was adopted. Figure 5.34 shows a photograph of the arrangement of the vibration experiment.

The fatigue limit of the specimen is tested using the up-and-down method. When the excitation frequency drops cumulatively to 3 Hz below the cycle index number 10^7, the test will stop and the specimen block can be defined as "Break"; when the vibration frequency decreases cumulatively below the value of 3 Hz, with the number 10^7 of the cycle index, the test will stop and the specimen block can be defined as "exceeded". When the proof stress of the specimen block exceeds a given value of 10% with the cycle index number 25%, the test will stop and if a crack appears on the gripping position of the specimen block, the test will be canceled. Figure 5.33 shows a photograph of the arrangement of the vibration experiment. The exciter provides vibration displacement at the resonance frequency of the sample mounted on it. The displacement and stress are monitored by a displacement sensor and a strain gauge. Strain gauges should be used to establish test stress levels by calibrating to the tip deflection. Gauges should be placed at areas of interest and reference locations on the part. If the highest stressed areas cannot be strain gauged, strain gauges should be used in conjunction with a model of a finite element to predict peak stresses. Before the test, the relation between the displacement at the tip of the sample and the maximum stress is determined. Under the room temperature atmospheric condition, the vibration fatigue of the K417 sample test is shown in Fig. 5.35.

Fig. 5.35 Clamp and measurement

Fig. 5.36 The fluctuating of the vibration fatigue strength of the K417 alloy under different processing statuses

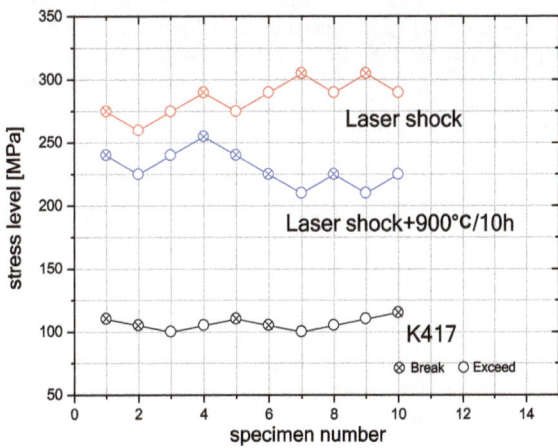

As shown in Fig. 5.36, compared with the original sample, the LSP can effectively increase the vibration fatigue strength from 110 to 285 MPa. After the heat treatment of 900 °C/10 h, the strength of the vibration fatigue is 230 MPa, slightly lower than what it was before. Heat treatment does not obviously reduce the strengthening effectiveness of LSP.

References

1. H.B. Long, S.C. Mao, Y.N. Liu, Z. Zhang, X.D. Han, Microstructural and compositional design of Ni-based single crystalline superalloysd—a review. J. Alloy. Compd. **743**, 203–220 (2018)
2. E. Chauvet, P. Kontis, E.A. Jagle, B. Gault, D. Raabe, C. Tassin, J.J. Blandin, R. Dendievel, B. Vayre, S. Abed, G. Martin, Hot cracking mechanism affecting a non-weldable Ni-based superalloy produced by selective electron beam melting. Acta Mater. **142**, 82–94 (2018)

3. M. Kattoura, S.R. Mannava, D. Qian, V.K. Vasudevan, Effect of laser shock peening on residual stress, microstructure and fatigue behavior of ATI 718Plus alloy. Int. J. Fatigue **102**, 121–134 (2017)
4. M. Kattoura, S.R. Mannava, D. Qian, V.K. Vasudevan, Effect of laser shock peening on elevated temperature residual stress, microstructure and fatigue behavior of ATI 718Plus alloy. Int. J. Fatigue **104**, 366–378 (2017)
5. C. Cellard, D. Retraint, M. François, E. Rouhaud, D. Le Saunier, Laser shock peening of Ti-17 titanium alloy: influence of process parameters. Mater. Sci. Eng., A **532**, 362–372 (2012)
6. S. Sathyajith, S. Kalainathan, S. Swaroop, Laser peening without coating on aluminum alloy Al-6061-T6 using low energy Nd:YAG laser. Opt. Laser Technol. **45**, 389–394 (2013)
7. C. Rubio-González, J.L. Ocaña, G. Gomez-Rosas, C. Molpeceres, M. Paredes, A. Banderas, J. Porro, M. Morales, Effect of laser shock processing on fatigue crack growth and fracture toughness of 6061–T6 aluminum alloy. Mater. Sci. Eng., A **386**(1), 291–295 (2004)
8. E. Maawad, Y. Sano, L. Wagner, H.G. Brokmeier, C. Genzel, Investigation of laser shock peening effects on residual stress state and fatigue performance of titanium alloys. Mater. Sci. Eng. A **536**, 82–91 (2012)
9. Z. Zhou, A.S. Gill, A. Telang, S.R. Mannava, K. Langer, V.K. Vasudevan, D. Qian, Experimental and finite element simulation study of thermal relaxation of residual stresses in laser shock peened IN718 SPF superalloy. Exp. Mech. **54**(9), 1597–1611 (2014)
10. Y. Li, L. Zhou, W. He, G. He, X. Wang, X. Nie, B. Wang, S. Luo, Y. Li, The strengthening mechanism of a nickel-based alloy after laser shock processing at high temperatures. Sci. Technol. Adv. Mater. **14**(5), 5010 (2013)
11. W.F. Zhou, X.D. Ren, Y.P. Ren, S.D. Xu, J.J. Huang, T. Yang, Laser shock processing on Ni-based superalloy K417 and its effect on thermal relaxation of residual stress. Int. J. Adv. Manuf. Technol. **88**(1–4), 675–681 (2017)
12. I. Nikitin, I. Altenberger, Comparison of the fatigue behavior and residual stress stability of laser-shock peened and deep rolled austenitic stainless steel AISI 304 in the temperature range 25–600 °C. Mater. Sci. Eng. A (2007)
13. N.F. Ren, H.M. Yang, S.Q. Yuan, Y. Wang, S.X. Tang, L.M. Zheng, X.D. Ren, F.Z. Dai, High temperature mechanical properties and surface fatigue behavior improving of steel alloy via laser shock peening. Mater. Des. **53**, 452–456 (2014)
14. Y. Yang, K. Zhou, G. Li, Surface gradient microstructural characteristics and evolution mechanism of 2195 aluminum lithium alloy induced by laser shock peening. Opt. Laser Technol. **109**, 1–7 (2019)
15. D. Karthik, K.U. Yazar, A. Bisht, S. Swaroop, C. Srivastava, S. Suwas, Gradient plastic strain accommodation and nanotwinning in multi-pass laser shock peened 321 steel. Appl. Surf. Sci. **487**, 426–432 (2019)
16. B. Mao, Y.L. Liao, B. Li, Gradient twinning microstructure generated by laser shock peening in an AZ31B magnesium alloy. Appl. Surf. Sci. **457**, 342–351 (2018)
17. A. Umapathi, S. Swaroop, Phase gradient in a laser peened TC6 titanium alloy analyzed using synchrotron radiation. Mater. Charact. **131**, 431–439 (2017)
18. R. Huang, Y. Han, Structure evolution and thermal stability of SMAT-derived nanograined layer on Ti-25Nb-3Mo-3Zr-2Sn alloy at elevated temperatures. J. Alloy. Compd. **554**, 1–11 (2013)
19. I. Altenberger, E.A. Stach, G. Liu, R.K. Nalla, R.O. Ritchie, An in situ transmission electron microscope study of the thermal stability of near-surface microstructures induced by deep rolling and laser-shock peening. Scripta Mater. **48**(12), 1593–1598 (2003)
20. W.J. Jia, H.Z. Zhao, Q. Hong, L. Li, X.N. Mao, Research on the thermal stability of a near alpha titanium alloy before and after laser shock peening. Mater. Charact. **117**, 30–34 (2016)
21. J.M. Yang, Y.C. Her, N.L. Han, A. Clauer, Laser shock peening on fatigue behavior of 2024–T3 Al alloy with fastener holes and stopholes. Mater. Sci. Eng. A-Struct. Mater. Prop. Microstruct. Process. **298**(1–2), 296–299 (2001)
22. S.H. Luo, L.C. Zhou, X.D. Wang, X. Cao, X.F. Nie, W.F. He, Surface nanocrystallization and amorphization of dual-phase TC11 titanium alloys under laser induced ultrahigh strain-rate plastic deformation. Materials **11**(4) (2018)

23. S.H. Luo, W.F. He, L.C. Zhou, X.F. Nie, Y.H. Li, Aluminizing mechanism on a nickel-based alloy with surface nanostructure produced by laser shock peening and its effect on fatigue strength. Surf. Coat. Technol. **342**, 29–36 (2018)

24. L. Chen, X.D. Ren, W.F. Zhou, Z.P. Tong, S. Adu-Gyamfi, Y.X. Ye, Y.P. Ren, Evolution of microstructure and grain refinement mechanism of pure nickel induced by laser shock peening. Mater. Sci. Eng. A-Struct. Mater. Prop. Microstruct. Process. **728**, 20–29 (2018)

25. M.A. Meyers, A. Mishra, D.J. Benson, Mechanical properties of nanostructured materials. Prog. Mater Sci. **51**(4), 427–556 (2006)

26. M.A. Meyers, F. Gregori, B.K. Kad, M.S. Schneider, D.H. Kalantar, B.A. Remington, G. Ravichandran, T. Boehly, J.S. Wark, Laser-induced shock compression of monocrystalline copper: characterization and analysis. Acta. Mater. **51**, 1211–1228 (2003).

27. I.I. Oleynik, B.J. Demaske, V.V. Zhakhovsky, N.A. Inogamov, C.T. White, MD simulations of laser-induced ultrashort shock waves in nickel (2011)

28. C. Huang, T.G. Murthy, M.R. Shankar, R. M'Saoubi, S. Chandrasekar, Temperature rise in severe plastic deformation of titanium at small strain-rates. Scripta Mater. **58**(8), 663–666 (2008)

29. I. Zarudi, T. Nguyen, L.C. Zhang, Effect of temperature and stress on plastic deformation in monocrystalline silicon induced by scratching. Appl. Phys. Lett. **86**(1), 437 (2005)

30. Y. Yang, H.M. Wang, K. Zhou, G.J. Li, Effect of laser shock peening and annealing temperatures on stability of AA2195 alloy near-surface microstructure. Opt. Laser Technol. **119** (2019)

31. A. Telang, T. Gnaupel-Herold, A. Gill, V.K. Vasudevan, Effect of applied stress and temperature on residual stresses induced by peening surface treatments in alloy 600. J. Mater. Eng. Perform. **27**(6), 2796–2804 (2018)

32. A. Telang, A.S. Gili, S.R. Mannava, D. Qian, V.K. Vasudevan, Effect of temperature on microstructure and residual stresses induced by surface treatments in Inconel 718 SPF. Surf. Coat. Technol. **344**, 93–101 (2018)

33. X.D. Hou, N.M. Jennett, Application of a modified slip-distance theory to the indentation of single-crystal and polycrystalline copper to model the interactions between indentation size and structure size effects. Acta Mater. **60**(10), 4128–4135 (2012)

34. Y.W. Bao, W. Wang, Y.C. Zhou, Investigation of the relationship between elastic modulus and hardness based on depth-sensing indentation measurements. Acta Mater. **52**(18), 5397–5404 (2004)

35. D.L. Joslin, W.C. Oliver, A new method for analyzing data from continuous depth-sensing microindentation tests. J. Mater. Res. **5**(1), 123–126 (1990)

36. L.C. Zhou, C.B. Long, W.F. He, L. Tian, W.T. Jia, Improvement of high-temperature fatigue performance in the nickel-based alloy by LSP-induced surface nanocrystallization. J. Alloy. Compd. **744**, 156–164 (2018)

37. Z.P. Tong, X.D. Ren, Y.P. Ren, F.Z. Dai, Y.X. Ye, W.F. Zhou, L. Chen, Z. Ye, Effect of laser shock peening on microstructure and hot corrosion of TC11 alloy. Surf. Coat. Technol. **335**, 32–40 (2018)

38. S.H. Luo, X.F. Nie, L.C. Zhou, X. You, W.F. He, Y.H. Li, Thermal stability of surface nanostructure produced by laser shock peening in a Ni-based superalloy. Surf. Coat. Technol. **311**, 337–343 (2017)

39. W.Q. Zhang, J.Z. Lu, K.Y. Luo, Residual stress distribution and microstructure at a laser spot of AISI 304 stainless steel subjected to different laser shock peening impacts. Metals **6**(1) (2016)

40. Y. Shadangi, K. Chattopadhyay, S.B. Rai, V. Singh, Effect of LASER shock peening on microstructure, mechanical properties and corrosion behavior of interstitial free steel. Surf. Coat. Technol. **280**, 216–224 (2015)

41. X.F. Nie, W.F. He, Q.P. Li, N.D. Long, Y. Chai, Experiment investigation on microstructure and mechanical properties of TC17 titanium alloy treated by laser shock peening with different laser fluence. J. Laser Appl. **25**(4) (2013)

42. R.J. Sun, L.H. Li, Y. Zhu, W. Guo, P. Peng, B.Q. Cong, J.F. Sun, Z.G. Che, B. Li, C. Guo, L. Liu, Microstructure, residual stress and tensile properties control of wire-arc additive manufactured 2319 aluminum alloy with laser shock peening. J. Alloy. Compd. **747**, 255–265 (2018)

43. M.Z. Ge, J.Y. Xiang, Effect of laser shock peening on microstructure and fatigue crack growth rate of AZ31B magnesium alloy. J. Alloy. Compd. **680**, 544–552 (2016)
44. S.H. Luo, Y.H. Li, L.C. Zhou, X.F. Nie, G.Y. He, Y.Q. Li, W.F. He, Surface nanocrystallization of metallic alloys with different stacking fault energy induced by laser shock processing. Mater. Des. **104**, 320–326 (2016)
45. M.A. Meyers, *Dynamic Behavior of Materials* (Wiley, NewYork, USA, 1994).
46. M. Lewandowska, K.J. Kurzydlowski, Thermal stability of a nanostructured aluminium alloy. Mater. Charact. **55**(4–5), 395–401 (2005)

Chapter 6
Mechanical Behavior and the Strengthening Mechanism of LSP-Induced Gradient Microstructure in Metal Materials

6.1 Introduction

In most cases, fatigue cracks are likely to initiate in the surface of metallic components. Therefore, optimization of surface microstructures and properties can effectively improve the reliability of parts and prolong the service lifetime of components [1, 2]. Due to the ultrafine grains and the large amounts of grain boundaries, nanocrystalline materials have been considered to exhibit many beneficial properties relative to the coarse grain materials [3, 4]. Therefore, many kinds of methods have been developed to synthetize nanocrystalline layer in the materials surface. Such surface nanocrystallization can be realized by the means of severe plastic deformation (SPD) [5]. LSP is one of the most common SPD methods to produce surface nanocrystalline layer without changing its chemical compositions [6]. Surface nanocrystallization has already been realized by LSP on many metals and alloys, such as Ni-based superalloys, titanium alloys, magnesium alloy and stainless steel [4, 7–9]. So far, high-value compressive residual stress and microstructure changes induced by SPD have been regarded as two main mechanisms for improving fatigue properties at room temperature [10–12]. However, these mechanisms are only effective in service when they remain stable under the cyclic and/or thermal loading. Therefore, this section mainly introduces three parts of research work. First of all, by establishing a cross-scale mechanical model of LSP-induced gradient nanostructure in titanium alloy, the strengthening-toughening mechanism of LSP-induced gradient structure is revealed; second, by establishing a molecular dynamics model of crack propagation, the microscopic mechanism of crack propagation in titanium alloy and the mechanical behavior of nanocrystals under tensile loading are revealed; third, the effects of LSP on the improvement of vibration fatigue performance of thin-walled components are also discussed and given.

© Zhejiang University Press 2021
L. Zhou and W. He, *Gradient Microstructure in Laser Shock Peened Materials*, Springer Series in Materials Science 314,
https://doi.org/10.1007/978-981-16-1747-8_6

6.2 Mechanical Behavior of the LSP-Induced Gradient Microstructure in Titanium Alloy

6.2.1 The Model of Crystal Plasticity of the LSP-Induced Gradient Microstructure in Titanium Alloy

In this section, TC4 is taken as the research object, and its chemical composition and mechanical properties are shown in Sect. 6.3.1. At room temperature, the TC4 alloy contains a 95% α-phase HCP structure and a 5% β-phase bcc structure. When the temperature reaches about 1163 K, α phase begins to transform into β phase. The results of current research show that the mechanical properties of the TC4 alloy will not be affected when the content of β phase is no more than 50%. Therefore, the influence of β phase is not considered in this section.

Gradient microstructure is introduced into the surface layer of the TC4 alloy by LSP. In this section, an elastic-viscoplastic constitutive model considering the effect of the grain size is established to study the mechanical behavior of gradient structures.

For elastic-viscoplastic solid materials, the total rate of strain can be divided into elastic and plastic parts:

$$\dot{\varepsilon}_{ij} = \dot{\varepsilon}_{ij}^{e} + \dot{\varepsilon}_{ij}^{p} \tag{6.1}$$

The relationship between the rate of elastic stress and the rate of elastic strain is described by Hooke's law:

$$\dot{\varepsilon}_{ij}^{e} = \frac{1}{2\mu}\dot{\sigma}_{ij}' + \frac{\dot{\sigma}_{kk}}{9K}\delta_{ij} \tag{6.2}$$

where μ and k are the elastic modulus and the bulk modulus respectively; $\dot{\sigma}_{ij}' = \dot{\sigma}_{ij} - \dot{\sigma}_{kk}\delta_{ij}/3$ is the rate of deviatoric stress, $\dot{\sigma}_{kk}$ is the rate of hydrostatic pressure and δ_{ij} is the Kronecker symbol.

The relationship between the rate of plastic strain and deviatoric stress is described by the J_2 flow rule:

$$\dot{\varepsilon}_{ij}^{p} = \frac{3\dot{\varepsilon}^{p}}{2\sigma_{e}}\sigma_{ij}' \tag{6.3}$$

where $\dot{\varepsilon}^{p} = \sqrt{2\dot{\varepsilon}_{ij}^{p}\dot{\varepsilon}_{ij}^{p}/3}$ is the rate of equivalent plastic strain and $\sigma_{e} = \sqrt{3\sigma_{ij}'\sigma_{ij}'/2}$ is equivalent stress.

The relationship between the rate of equivalent plastic strain and that of equivalent strain is described by the power-law hardening relationship[13]:

$$\dot{\varepsilon}^{\mathrm{P}} = \dot{\varepsilon} \left(\frac{\sigma_{\mathrm{e}}}{\sigma_{\mathrm{f}}} \right)^{m} \tag{6.4}$$

where m is the sensitive coefficient, $\dot{\varepsilon} = \sqrt{2\dot{\varepsilon}'_{ij}\dot{\varepsilon}'_{ij}/3}$ is the equivalent strain, $\dot{\varepsilon}'_{ij} = \dot{\varepsilon}_{ij} - \dot{\varepsilon}_{kk}\delta_{ij}/3$ is the rate of deviatoric strain and σ_{f} is flow stress, which is determined by the microstructure and deformation mechanism of the material.

Twinning is an important deformation mechanism for HCP materials. However, the addition of aluminum can effectively inhibit twinning. Twinning is not an important deformation mechanism for the TC4 alloy. Therefore, in this section, only the dislocation mechanism is considered. The flow stress of the TC4 alloy can be divided into two parts [14, 15]: the heat activated part σ_{th} and the non-heat part σ_{ath}.

$$\sigma_{\mathrm{f}} = \sigma_{\mathrm{th}} + \sigma_{\mathrm{ath}} \tag{6.5}$$

where, the heat activated part is used to overcome lattice friction and short-range obstacles, such as solid-solution atoms, encountered by dislocation in the process of motion. The non-heat part is used to overcome long-range obstacles, such as the interaction among dislocations.

The dislocation overcoming a short-range obstacle can be assisted by temperature, which is a process of heat activation. According to the Orowan equation,

$$\dot{\gamma} = \rho_{\mathrm{m}}\bar{v}b \tag{6.6}$$

where $\dot{\gamma}$ is slip rate, ρ_{m} is the density of mobile dislocation, $\bar{v}$ is the average velocity of the dislocation movement, and b is the magnitude of Berg's vector. This model describing the slip rate of single crystal can be related by the Taylor factor M [16], and the relationship between the rate of macroscopic equivalent plastic strain and that of the equivalent slip is $\dot{\gamma} = M\dot{\varepsilon}^{\mathrm{P}}$, so

$$\dot{\varepsilon}^{\mathrm{P}} = \frac{\rho_{\mathrm{m}}\bar{v}b}{M} \tag{6.7}$$

The average velocity of the dislocation motion is:

$$\bar{v} = \frac{\lambda}{t} \tag{6.8}$$

where λ is the average distance between obstacles, t is the time needed for dislocations to cross obstacles, most of which are the waiting time of dislocations in front of obstacles, which can be described as:

$$t = \frac{1}{v_a} e^{\Delta G / k_{\mathrm{B}} T} \tag{6.9}$$

where v_a is the frequency of dislocation impacting obstacles, $e^{\Delta G/k_B T}$ is the probability of overcoming the obstacles, ΔG is the activation energy, K^B is the Boltzmann constant, and T is the temperature. Formulas (6.8) and (6.9) can be brought into Formula (6.7) to get:

$$\dot{\varepsilon}^p = \frac{\rho_m b}{M} \lambda v_a e^{-\Delta G/k_B T} \tag{6.10}$$

The activation energy of dislocation motion can be described by the following Formula [17]:

$$\Delta G = \Delta F \left[1 - \left(\frac{\sigma_{th}}{\sigma^*} \right)^p \right]^q \tag{6.11}$$

where $\Delta F = f_0 \mu b^3$ is the activation energy of dislocation to overcome obstacles, f_0 is the parameter that determines the strength of the obstacles, $\sigma^* = g_0 \mu$ is the shear stress without thermal assistance, g_0 is related to the strength of obstacles, and p and q are the parameters that describe the outline of obstacles. Through the simultaneous Eqs. (6.10) and (6.11), the final form σ_{ath} can be given as [18]:

$$\sigma_{th} = g_0 \mu \left[1 - \left[\frac{k_B T}{f_0 \mu b^3} \ln \left(\frac{\dot{\varepsilon}^{ref}}{\dot{\varepsilon}^p} \right) \right]^{1/q} \right]^{1/p} \tag{6.12}$$

where $\dot{\varepsilon}^{ref}$ is the reference rate of the plastic strain.

The thermal stress used to overcome the interaction between grain boundaries and dislocations can be described by the following formula:

$$\sigma_{ath} = \frac{k_{HP}}{\sqrt{d}} + M \alpha \mu b \sqrt{\rho} \tag{6.13}$$

where $k_{HP}/\sqrt{d}$ represents the action from grain boundaries [11], k_{HP} is the Hall–Petch slope, and d is the grain size; the Taylor hardening rate $M \alpha \mu b \sqrt{\rho}$ indicates tree dislocation strengthening [19, 20], where M is the Taylor factor, α is the material constant and ρ is total dislocation density.

In the deformation process of the TC4 alloy strengthened by LSP, gradient microstructure will introduce the gradient deformation in the deformation process. Therefore, the total dislocation density is further divided into two parts: geometrically necessary dislocations and statistical storage dislocations.

$$\rho = \rho_{SSDs} + \rho_{GNDs} \tag{6.14}$$

The relationship between density of the geometrically necessary dislocation and the gradient plastic deformation is [21, 22]:

$$\rho_{\text{GNDs}} = \bar{r}\frac{\eta^{\text{p}}}{b} \tag{6.15}$$

where $\bar{r}$ is the Nye factor, η^{p} can be calculated as [23, 24]:

$$\eta^{\text{p}} = \sqrt{\frac{1}{4}\eta^{\text{p}}_{ijk}\eta^{\text{p}}_{ijk}} \tag{6.16}$$

where $\eta^{\text{p}}_{ijk} = \varepsilon^{\text{p}}_{ik,j} + \varepsilon^{\text{p}}_{jk,i} - \varepsilon^{\text{p}}_{ij,k}$. The gradient plastic strain can be solved by user subroutine, see [25] for details. The density of statistical storage dislocation is described by the modified KME model [26]:

$$\frac{\partial \rho_{\text{SSDs}}}{\partial \varepsilon^{\text{p}}} = M\left[\frac{k_0}{bd} + \frac{k_1}{b}\sqrt{\rho} - k_2\left(\frac{\dot{\varepsilon}^{\text{p}}}{\dot{\varepsilon}_0}\right)^{-\frac{1}{n_0}}\rho_{\text{SSDs}} - \left(\frac{d_e}{d}\right)^2\rho_{\text{SSDs}}\right] \tag{6.17}$$

where, k_0 and k_1 are geometric factors related to grain size and dislocation density, k_2 and $\dot{\varepsilon}_0$ are material constants, n_0 is related to temperature, and d_e is the reference grain size, which is used to describe the fact that the annihilation of dislocation at grain boundaries becomes stronger when it is smaller than this size. The second and third items in brackets describe dislocation increment and dislocation annihilation in the original KME model. In the modified KME model, the first item in brackets describes dislocation increment caused by the decrease in dislocation's mean free path caused by the existence of grain boundaries, and the fourth item describes the annihilation effect of dislocation at grain boundaries.

The above establishes a constitutive model describing the effect of grain size and the deformation behavior of the TC4 alloy, and can be used to simulate the deformation behavior of the TC4 alloy subjected to LSP with gradient microstructure.

6.2.2 Multi-scale Mechanical Behavior of the LSP-Induced Gradient Microstructure

In this book, 1/4 model is established for simulation, as shown in Fig. 6.1, and the distribution of the grain size is shown in Table 6.1. According to the experimental setup, uniaxial tension is applied in the x direction, the strain rate is 10^{-3}/s, and symmetric constraints are applied on the x–y, x–z and negative y–z planes.

In the process of LSP, gradient microstructure and residual stress are introduced simultaneously. The above method can simulate the structure of the gradient grain, and the residual stress can be considered by applying a prestress field in the finite element simulation. There are two methods to simulate the residual stress field, namely, directly simulating the LSP process or using the thermal engine method [27–30]. For the former, the process of LSP is very complicated and time-consuming.

Fig. 6.1 Finite element
model of the TC4 specimen
strengthened by LSP

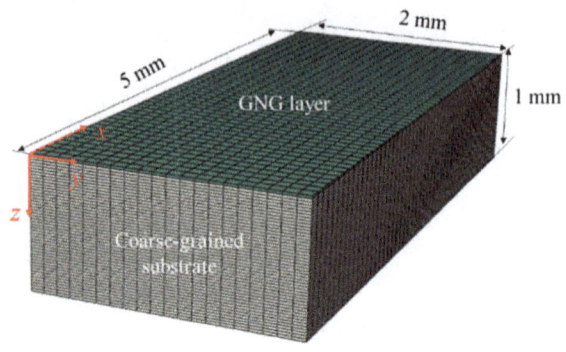

Table 6.1 Grain size
distribution

Depth (μm)	Grain size (μm)
0–20	1
20–1000	5

Relatively, the latter can easily obtain a residual stress field consistent with the
experimental results, which is convenient to operate and has no requirement for
the mesh of finite element simulation [31]. Therefore, this section uses this method
to obtain the residual stress field. As shown in Fig. 6.2, the stress field measured
by the experiment is only residual compressive stress, but not tensile stress, which
will lead to the imbalance of force in the sample. Relatively, the residual stress field
obtained by the finite element method can ensure a balance of force in the sample.
The layer-by-layer stripping method has little effect on the residual stress of the
surface layer, but has a great influence on the core. Here, the simulation results of
the finite element only correspond to the experiments in the surface area, while the
field of balanced force obtained by the finite element simulation is maintained in the
core, as shown in Fig. 6.2.

Fig. 6.2 Comparison of the
residual stress of the finite
element simulation and
experimental results

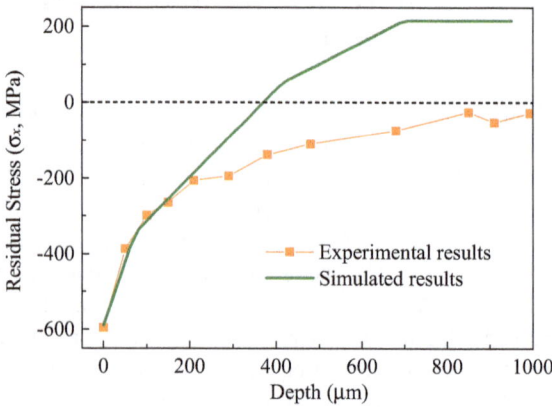

Using the constitutive model, the method of finite element simulation and residual stress established above, the influence of residual stress and gradient microstructure can be considered at the same time, and their effects can also be decoupled for a detailed study.

In addition, in order to study the influence of gradient structure on tensile properties, quasi-static uniaxial tensile tests are carried out on the original TC4 alloy and LSPed TC4 alloy at room temperature using MTS-858 tensile and a torsional testing machine. In order to study the effect of LSP on elastic modulus, initial yield and subsequent hardening, all specimens are stretched to a strain of about 4% and a strain rate of 10^{-3}/s. Strain is measured by means of an extensometer. The true stress and strain curves of uniaxial tensile are shown in Fig. 6.3. It can be seen that the LSPed TC4 alloy has a longer elastic–plastic transition section, and the stress–strain curve of the LSPed TC4 alloy is separated from the curve of the original sample from the strain of 0.005. Therefore, the elastic modulus of the LSPed samples decreases. However, it should be noted that the elastic modulus of the LSPed sample does not change, because the stress–strain curves of the LSPed and untreated samples coincide at the initial stage of deformation. Therefore, the separation of the two curves only indicates that the LSPed specimen enters into plastic deformation earlier. Because the curve in Fig. 6.3 has no obvious yield point, it is analyzed that the yield stress is 0.2% here. It can be seen that the yield stresses of the original and LSPed samples are 834.3 MPa and 782.6 MPa, respectively. The yield stress of the material is reduced by 52 MPa by means of LSP. In the subsequent hardening stage, the stress of the LSPed material gradually exceeds that of the untreated material ($\varepsilon > 0.012$). Finally, the stress of the strengthened material is about 30 MPa higher than that of the original material.

In order to study the reasons for premature yield, a decrease in yield stress and an increase in the strength in the subsequent hardening stage, a constitutive model based on the deformation mechanism is established to study the mechanism of influence of gradient microstructure in detail.

Fig. 6.3 Single tensile stress–strain curves of untreated and LSPed materials

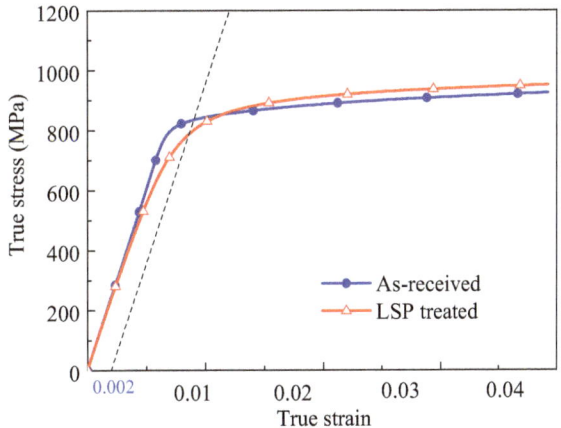

Gradient structures with different grain sizes are formed in the TC4 alloy by LSP. In order to verify the validity of the model and obtain the model parameters, first, the uniform grain structure TC4 with different grain sizes is simulated. The simulation results of the TC4 alloy with a grain size of 5 μm, 2 μm, 1 μm and 500 nm are given in Fig. 6.4. The experimental results with a grain size of 5 μm and 2 um are also shown in the figure for comparison [32]. It can be seen that the results of the simulation can correspond well with the experimental results, which shows that the established model combined with parameters can well predict the tensile mechanical properties of TC4 alloys with different grain sizes.

Figure 6.5 shows the results of simulating the uniaxial tensile response of the TC4 alloy treated by LSP using the model established above and the parameters obtained, and the results of the simulation in Fig. 6.3 are used for residual stress. The experimental and simulation results of untreated materials are also compared in the figure.

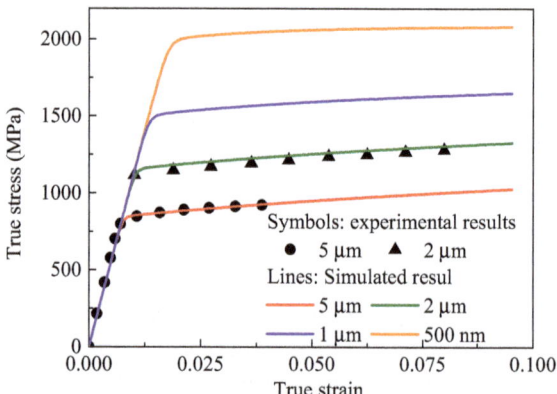

Fig. 6.4 Uniaxial tensile stress–strain response of TC4 alloy with a uniform grain size through constitutive modeling and its comparison with experimental results

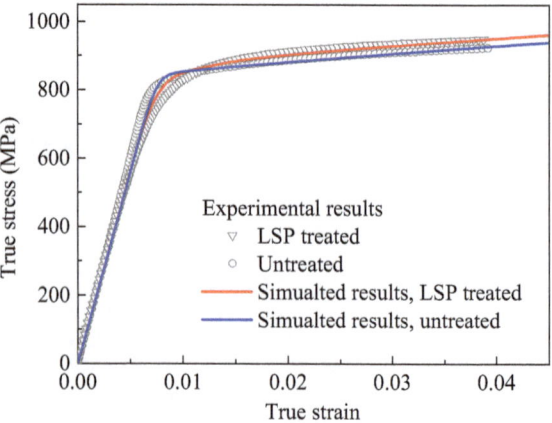

Fig. 6.5 Results of the simulation of LSPed and untreated materials and their comparison with experimental results

Table 6.2 Assumed grain size distribution in gradient layers

Depth (μm)	Grain size (μm)
0–20	0.3
20–40	0.4
40–60	0.5
60–80	0.8
80–120	1
120–140	2
140–1000	5

It can be seen from Fig. 6.5 that the results of the simulation can correspond well with the experimental results. Compared with the untreated samples, the premature yield and the increase in strength in the subsequent strain hardening stage of the LSPed samples are well predicted. However, the factors affecting these changes are still unclear. The effects of gradient microstructure and residual stress will be studied separately below.

The influence of gradient microstructure can be analyzed by comparing the responses of coarse grain and gradient grain structure. Because the gradient layer produced by LSP is very thin in this experiment, it is not convenient to analyze the influence of gradient microstructure. Therefore, a deeper gradient layer is set here, as shown in Table 6.2.

Figure 6.6 shows the results of the simulation of coarse grain and gradient grain TC4 alloy. It can be seen that the existence of a gradient layer will affect the initial yield and subsequent hardening of materials at the same time. This is consistent with the results of copper [33, 34] and interstitial-free atomic steel [35] with a gradient grain structure. It should be noted that the elastic limit of the material has not changed, the 0.2% yield stress is increased by 30 MPa, and the strength in the subsequent hardening stage is increased by 120 MPa. Some studies have shown that geometrically

Fig. 6.6 Results of the simulation of the TC4 alloy with coarse and gradient grain structure

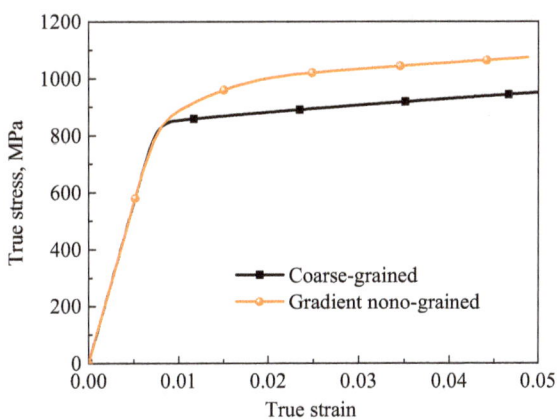

Dislocation density ($\times 10^{12}/m^2$)

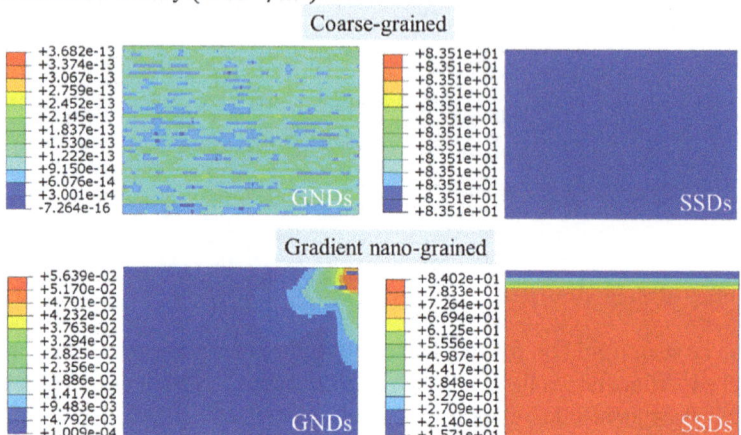

Fig. 6.7 Distribution of geometrically necessary dislocations and statistical storage dislocations in a TC4 alloy with coarse and gradient grain structures

necessary dislocations during the deformation of materials with a gradient grain structure can also lead to additional strengthening [23]. However, there is no obvious evidence to prove this. In order to study the contribution of geometrically necessary dislocations to strain hardening of the TC4 alloy treated by LSP, the size and distribution of these dislocations are analyzed in detail below.

Figure 6.7 shows the distribution of geometrically necessary dislocations and statistically stored dislocations under the condition that strain is 0.0488. It can be seen that the geometrically necessary density of dislocation in coarse grains is 0. In the gradient grain structure, the gradual yielding from the coarse grain core to the surface layer leads to non-uniform deformation and introduces geometrically necessary dislocations. Geometrically necessary dislocations are mainly distributed in the transition region between coarse grain and gradient layer, and their density is of the magnitude of $10^{10}/m^2$, which is far lower than the statistical storage dislocation density. Therefore, although geometrically necessary dislocations are introduced into the gradient grain structure, their density is very low and their area of distribution is limited, which makes no contribution to the strengthening of the TC4 alloy with a gradient grain structure. This is consistent with the conclusion of the strain gradient theory [36–39], that is, the effect of strain gradient only begins to contribute to the strengthening of materials when the density of geometrically necessary dislocation is equal to the density of statistical storage dislocation, such as the case where materials are subjected to non-uniform deformation at micro-scale [40].

Statistics show that the density of dislocation increases from $1.57 \times 10^{13}/m^2$ in the surface layer to $8.4 \times 10^{13}/m^2$ in the core layer, which is due to the smaller grain size in the surface layer and the stronger absorption of dislocation at the grain boundary. In coarse grains, the density of the dislocation distribution is uniform. Although the average density of dislocation in coarse grain is higher than that in a TC4 alloy with a gradient grain structure, the contribution of the density of dislocation to strengthening, namely $f_{\text{Hardening}}\left(\Delta\rho^{1/2}\right)$, is smaller than that of grain size to strength, namely $f_{\text{Yielding}}\left(\Delta d^{-1/2}\right)$. Therefore, the improvement in the strength of the TC4 alloy with gradient grain structure is mainly contributed by grain refinement.

6.3 A Molecular Dynamics Simulation of Crack Propagation in Pure Titanium Under Uniaxial Tension After LSP

6.3.1 A Molecular Dynamics Simulation of Crack Propagation in Pure Titanium Under Uniaxial Tension

Gradient structure formed by LSP can effectively improve the fatigue performance of metal parts. However, gradient structure involves multi-scale microstructure characteristics such as nanocrystals and dislocations, and the micro-mechanism of crack formation and propagation has always been a hot research issue. However, crack propagation is often completed instantaneously, and it is difficult to observe the law of the evolution of microstructure in the process of crack propagation by existing means of experimentation. Molecular dynamics (MD) simulation is a micro-scale simulation method. Because of its time and space scale characteristics, MD simulation is very suitable for studying the micro-deformation mechanism of crack propagation in metal materials. Therefore, this book takes titanium, which is widely used in the aviation field, as an example, and studies the mechanism for micro-plastic deformation of crack propagation of pure titanium and nano-polycrystalline titanium subjected to LSPby means of molecular dynamics simulation and experimental observation [41].

The model, based on molecular dynamics and used for modeling crack growth in pure titanium, is 10.24 nm $\times$ 43.05 nm $\times$ 44.34 nm in size and consists of totally 1,078,400 atoms. And the lattice parameters of Ti are a = 2.95 Å and c = 4.68 Å, respectively. The model is loaded using piston method (the piston is made of atomic layer with a thickness of 1 nm from top to bottom), and is stretched in the X direction. Y direction and Z direction adopt periodic boundary condition, while Y direction (in which crack grows) adopts free boundary condition. The X-, Y-, and Z-axes correspond to the [2-1-10], [-12-10], and [0001] orientations, which are shown in Fig. 6.1. Before being loaded, the model undergoes a thermodynamics relaxation under constant temperature using conjugate gradient method.

The uniaxial tension is applied with constant strain rate, in which the relation between size of loading direction and time is $L(t) = L_0(1 + erate \times dt)$. L_0 is the initial length of the sample, and *erate* is the strain parameter. An initial crack is introduced by deleting the atoms located on the central of X direction. The gap, with a length of 3 nm and width of 0.6 nm, is preset on the margin of the material, as shown in Fig. 6.8 (XY plane). In the simulation, the strain rate is set as

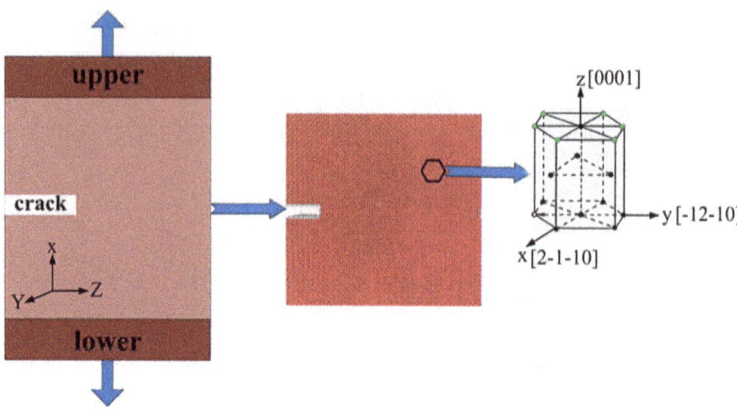

Fig. 6.8 The model and loading method of crack growth

2×10^9 s^{-1} and 4×10^9 s^{-1}, environment temperature as approximately 0, 300 and 600 K, and the initial pressure all zero. The purpose is to study the effects of strain rate and temperature on crack morphology, internal organization and mechanical property.

In this paper, all the atomic simulations are conducted in the open source MD programs large-scale atomic molecular massively parallel simulation (LAMMPS). Before loading, the relaxation lasts for 50,000 time steps, to get an initial equilibrated state, before the model comes to the steady state. Prior to stretch, the specimen are first equilibrated to achieve a minimum energy state, and the initial temperature is set as 300 K. During the relaxation process, it is based on Nose-Hover thermostat and stress (σ_{xx}) adjusts to about zero by baroatat in canonical ensemble (constant NPH). Ovito is employed to visualize the evolution of the atomic structure. In ovito, common neighbor analysis (CNA) and entrosymmetry parameter (CSP) are mainly used to analyze dislocation, stalking faults and concentration of stress.

Atom potential function plays a key role in MD simulation. Among numerous interatomic potentials, embedded atom potential uses local electron density as the key variable to describe the interaction among atoms in the metallic materials, including local interaction, like surface property and dislocation, and therefore it is suitable for the study of fracture and damage properties, the applicability of which has also been verified [42, 43]. The equations of motion are integrated using the Verlet leapfrog method with the time-step of 0.002 ps. The atom potential of titanium used in this study is Embedded Atom Method (EAM) potential developed by Sun et al. [44], which treats every atom as an impurity embedded into the system. Its total energy expression is:

$$\text{EAM:} E = \sum_i G_i \left(\sum_{j \neq i} \rho_j^a (r_{ij}) \right) + \frac{1}{2} \sum_{ij} U_{ij}(r_{ij}) \qquad (6.18)$$

In above expression, G_i stands for embedding energy, ρ_j^a stands for electron density where there is no atom i, U_{ij} stands for pair potential, γ_{ij} stands for the distance between atom i and atom j.

(1)   Crack growth mechanism

In this study, the micro-structure analysis of the crack growth in pure titanium is done through Common Neighbor Analysis (CNA). As shown in the Figure, red atoms stand for HCP structure, green atoms FCC structure, blue atoms BCC structure and grey atoms irregular structure. When the initial crack is preset with an initial temperature of 300 K and strain rate of 2×10^9 s^{-1}, the atomic arrangement around the crack in Fig. 6.9a becomes disordered, i.e. atoms marked by "1" are still of close-packed hexagonal structure, while atoms marked by "2", which are nearer the crack, are not arranged in regular hexagon, and the atomic arrangement is no longer of regular close-packed hexagonal structure. In the initial stretch stage, as crack grows to the top left area, the crack opening becomes bigger and bigger, and the crack tip gradually becomes disordered. At this time, crack tip atoms are rearranged, and stray from the ideal structure of lattice. Different from the initial atomic arrangement, the crack tip now becomes distorted. When ε is 0.002, the tiny voids are formed around crack tip clearly in Fig. 6.9b. According to the nucleation mechanism of voids, voids are formed when stress is greater than the binding force among atoms, which causes the atomic bond to fracture, or when massive vacancy clustering appears. Obviously, under the stretch, with the proceeding of loading, the plastic deformation is further enhanced. As shown in Fig. 6.9c, when strain ε is 0.01, the dislocation direction forms an angle of 45° relative to the surface. Local interatomic gap widens under the

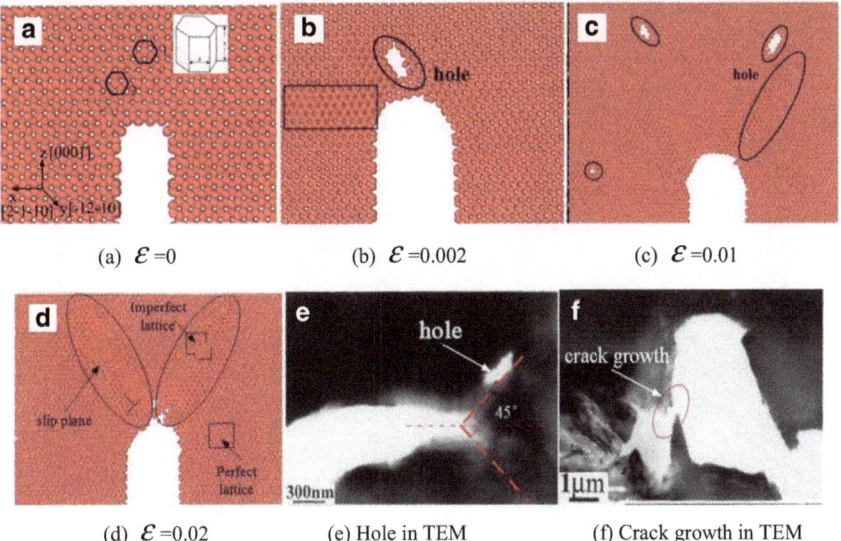

(a) $\varepsilon = 0$ (b) $\varepsilon = 0.002$ (c) $\varepsilon = 0.01$

(d) $\varepsilon = 0.02$ (e) Hole in TEM   (f) Crack growth in TEM

Fig. 6.9 The mechanism of crack growth

effects of force, and gradually becomes vacancy, which, after undergoing continuous development, finally forms into distinct voids. The random arrangement of atoms, as it were, inevitably leads to the formation of voids. With the further enhancement of strain, random arrangement of atoms and local stress concentration occur at a larger scale around crack tip. At this time, crack growth must overcome greater resistance, which results in a still more random arrangement and rearrangement of atoms around crack tip, thus sharpening the crack tip. After crack tip is sharpened, it expands swiftly in width and length, the void vanishes and becomes a part of crack and the crack opening becomes larger, expanding continuously in width and length, as shown in Fig. 6.9d. Experimental evidence for the hole and void vanishes is observed in α-Fe as well, the TEM observations were performed using a Philip H-800 TEM [45] and the TEM micrograph is clear in Fig. 6.9e, f.

Then the initial crack is preset in the form of a regular cuboid as shown in Fig. 6.10a, with the same initial temperature of 300 K and strain rate of 2×10^9 s^{-1}. Under the tension from both ends, when ε is 0.002, crack section becomes wedge-shaped, as shown in Fig. 6.10b. As the strain increases, vacancy keeps forming and gradually spreads to the voids. It is obvious the voids continuously absorb the vacancy

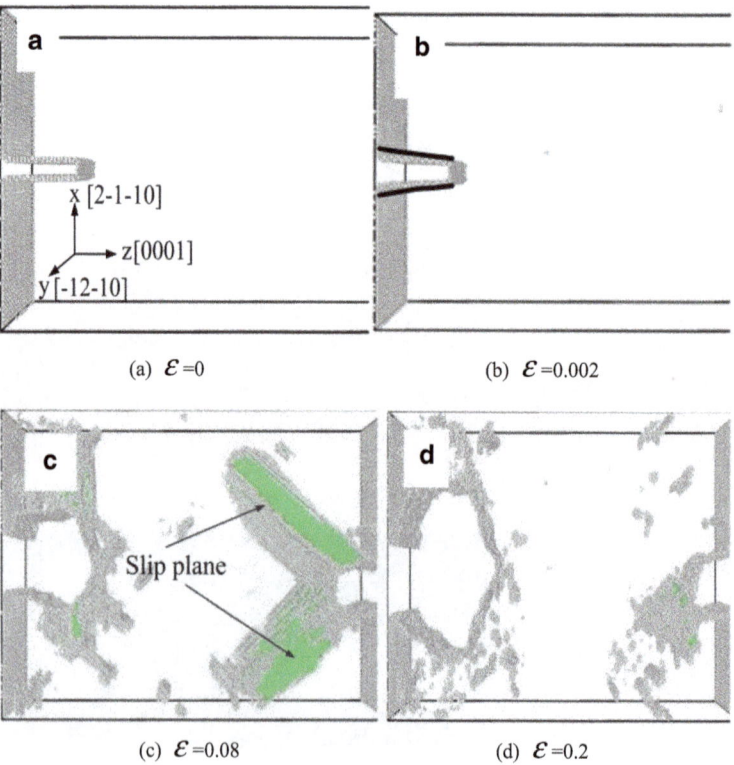

(a) $\mathcal{E}$=0   (b) $\mathcal{E}$=0.002   (c) $\mathcal{E}$=0.08   (d) $\mathcal{E}$=0.2

Fig. 6.10 The shape evolution of crack

and grow larger. Meanwhile, the atomic arrangement around the crack tips becomes more disordered, and takes on an irregular shape, as shown in Fig. 6.10c. When ε is 0.08, crack tip and surface are both uneven. The growth and clustering of countless voids, big or small, result in rip and new crack at the border of the right-hand area. The crack surface, under tension, becomes sunken towards the direction of tension, as shown in Fig. 6.10d.

In order to study the crack growth intensively, the microstructure deformation analyzed when the $\varepsilon = 0.6, 0.75, 1.5$. As shown in Fig. 6.11a, when ε is 0.6, phase transformation from HCP-BCC is observed, the BCC lattice occurs in crack tip, and the percentage of BCC in the sample is approximately 5%. Phase transformation from HCP to BCC is aroused by multiple "shockly" imperfect dislocation in the restriction of size dimension. As the increasing of strain, stalking faults was observed in sample, and the extension of stalking faults is stopped by twinning boundary, which are all observed in Fig. 6.11d. These stalking faults are located in sample symmetrically, and become fragmentization. When ε is 0.125, the stalking faults become an integral whole, and surrounded by twinning boundary. As shown in Fig. 6.11f. Experimental evidence for the stalking faults and twinning boundary is observed in HCP metal as well, the TEM observations were performed using a tenupol-3 electropolisher [45] and the TEM micrograph is clear in Fig. 6.11f.

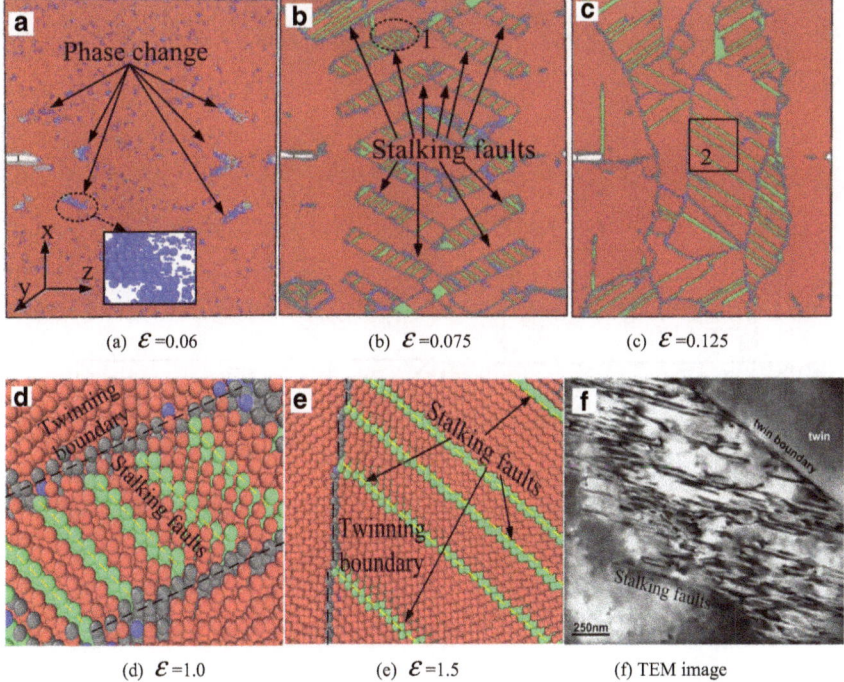

(a) $\mathcal{E}=0.06$ (b) $\mathcal{E}=0.075$ (c) $\mathcal{E}=0.125$

(d) $\mathcal{E}=1.0$ (e) $\mathcal{E}=1.5$ (f) TEM image

Fig. 6.11 The microstructure deformation of sample

(2)   Effect of strain rate and temperature on structural evolution

The plastic deformation of crack tip plays an important role in crack growth. It is plastic deformation that releases stress, which can blunt or sharpen crack. In order to study plastic deformation of crack tip, a stretching mode with uniaxial constant strain rate is adopted in the MD simulation, and the overall strain of the sample is designed to reach 0.2. Figure 6.12 shows the forming process of different forms of plastic deformation under different temperatures.

According to microcosmic failure mode of fatigue crack growth [46], crack tip is likely to develop into single shear form at low growth rate, and into double slip band at moderate growth rate. In the simulation, the environment temperature is 300 K, and the strain rate is $2 \times 10^9 \, \mathrm{s}^{-1}$. When the strain ε is 0.008, the crack propagates along the slip direction, the lattice begins to get distorted under the stress, phase transformation from HCP structure to BCC structure occurs at a small scale, crystals are sliding on the {0001} plane along [1120] and stacking-fault structures made of irregular atoms are formed along the aforesaid direction, as shown in Fig. 6.12a. When the strain ε increases to 0.075, the external force becomes the dominating factor for lattice deformation. Under the effects of force, phase transformation from HCP structure to FCC structure occurs at a large scale, forming into misfit dislocation network,

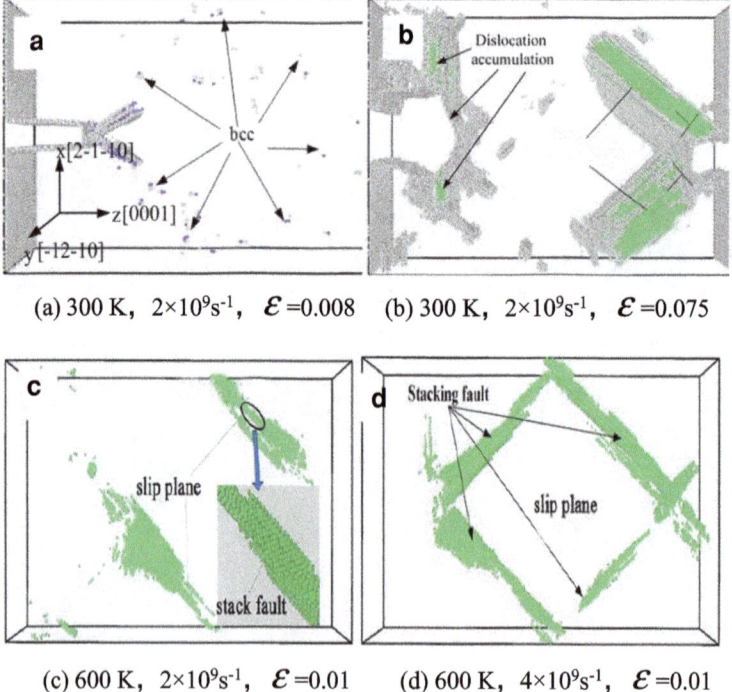

(a) 300 K, $2{\times}10^9\mathrm{s}^{-1}$, $\mathcal{E}$ =0.008   (b) 300 K, $2{\times}10^9\mathrm{s}^{-1}$, $\mathcal{E}$ =0.075

(c) 600 K, $2{\times}10^9\mathrm{s}^{-1}$, $\mathcal{E}$ =0.01   (d) 600 K, $4{\times}10^9\mathrm{s}^{-1}$, $\mathcal{E}$ =0.01

Fig. 6.12 The variation of microstructure

and stacking-fault structure is formed along the loading direction at the initial crack tip. Meanwhile, in the right-hand area, at the crack formed by rip, stacking fault can be clearly observed along {0001} slip plane and [1120] lattice direction, due to concentration of stress, which leads to dislocation accumulation, as shown in Fig. 6.12b. In order to study the effects of temperature and strain rate on stacking fault, the temperature is raised to 600 K, and strain rate to 4×10^9 s^{-1}. The results are as shown in Fig. 6.12c, d. To make the results clearer, here we only preserve stacking fault structure. When the temperature alone is raised, the strain decreased obviously during formation of stacking fault. Obviously, atomic activity is enhanced under high temperature, which facilitates formation of stacking fault. Strain rate also has a great impact on stacking fault. When the strain rate is 2×10^9 s^{-1}, the formation rates of stacking fault are along {0001} slip plane, but without symmetric growth of stacking fault structures. However, when the strain rate increases to 4×10^9 s^{-1}, symmetric growth of stacking fault structures can be observed both at initial crack and ripped crack tip on the {0001} slip plane along [1120] lattice direction.

In Fig. 6.13, strain ε is 0.05, temperature is 300 K, with the increase of strain rate, the sample occur different microstructure deformation obviously. When strain rate is 10^8 s^{-1}, there is no obvious microstructure deformation. However, phase transformation, stalking faults and the percentage of BCC increase with the increase of strain rate. The phenomenon can be observed clearly in Fig. 6.14, with the increase of temperature, phase transformation and defect atom number become more and more, dislocation nucleation become more easily. Compare with Fig. 6.14a, b, with the increase of temperature, the elastic modulus of pure titanium declines quickly, and the yield point of pure titanium drops with the increase of strain rate. It can draw a conclusion that: stalking fault has obvious influence on yield point, and phase transformation affects elastic modulus deeply.

(3) Initial crack effect

It is found in the experiment that titanium is very sensitive to crack, whose fatigue strength declines greatly in case of even tiny crack. In the simulation, a comparison

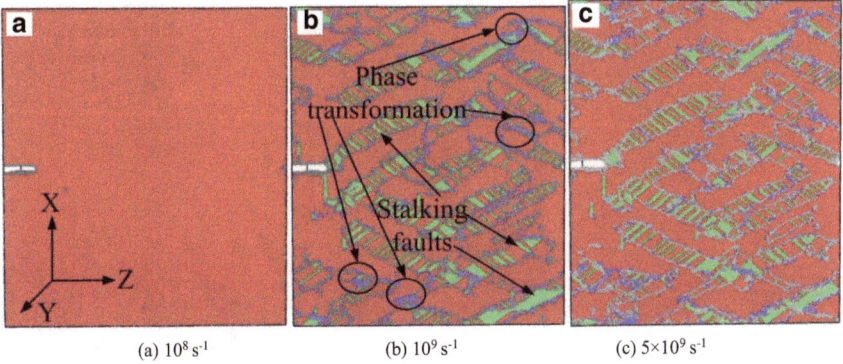

(a) 10^8 s^{-1} (b) 10^9 s^{-1} (c) 5×10^9 s^{-1}

Fig. 6.13 Effect of strain rate on microstructure deformation, $\varepsilon = 0.05$, 300 K

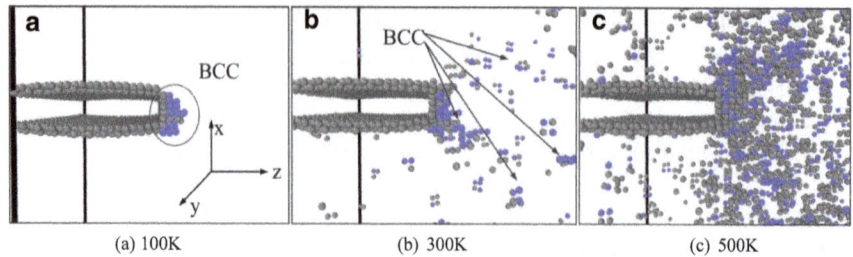

(a) 100K   (b) 300K   (c) 500K

Fig. 6.14 Effect of temperature on microstructure deformation, $\varepsilon = 0.05$, $\dot{\varepsilon} = 10^8$ s^{-1}

is made between stress–strain responses of pure titanium without initial crack and sample with tiny crack under tensile load. As shown in Fig. 6.15, in these cases, the temperature is 300 K and strain rate is 10^8 s^{-1}, the crack forms on the sample as 0, 5c, 10c, 20c (c = 4.68 Å), the yield point drops from 4.24 GPa to 3.28 GPa, and tension yield point drops by about 22.64%. Meanwhile, affected by cracks, the limiting strain in the stage of elastic deformation decreases a lot in the stress–strain curve. It can be observed in Fig. 6.15 that limiting strain of elastic stage in no-crack stress–strain curve is about 0.075, limiting strain in crack (20c) is about 0.05, it drops by about 33.33%. It demonstrates that crack size has significant impact on the carrying capacity of pure titanium. Figure 6.15 shows how crack is formed on pure titanium with no initial crack under tension. During the preliminary stage of tension, the plastically deforming area, which is characterized by stress concentration, gradually becomes vacancy, as shown in Fig. 6.16a. As strain increases, vacancy keeps gathering and forms into voids, as shown in Fig. 6.16b. When the strain ε is 0.03, the atomic disorder around voids becomes greater, and the voids keep growing in size, but the surface atoms are still connected, and no crack is formed, as shown in Fig. 6.16c. However, when the strain becomes 0.08, crack can be clearly observed, as shown in Fig. 6.16d. The crack, under the influence of original plastic accumulation, expands

Fig. 6.15 Influence of the crack size on the strain–stress response of pure titanium

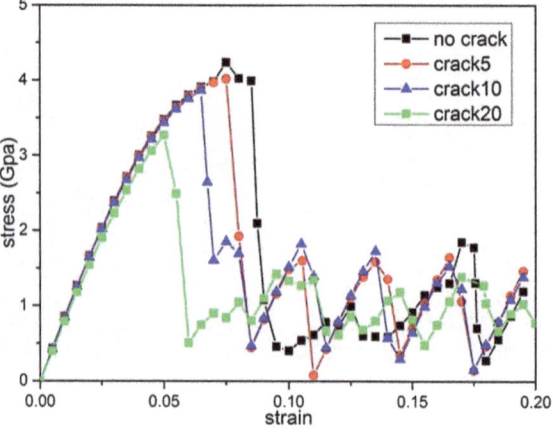

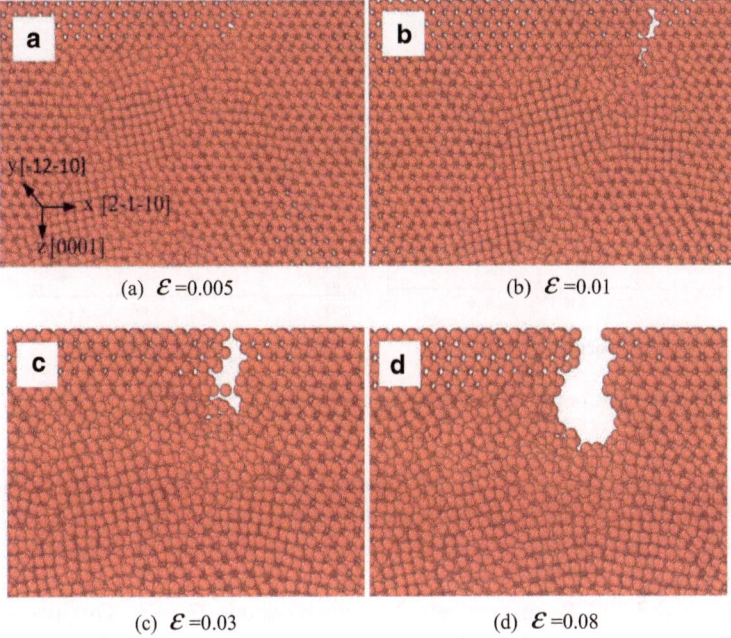

(a) $\mathcal{E}$=0.005   (b) $\mathcal{E}$=0.01

(c) $\mathcal{E}$=0.03   (d) $\mathcal{E}$=0.08

Fig. 6.16 The process of crack initiation of pure titanium (no initial crack), 300 K, 10^8 s^{-1}

rapidly, and the forming of crack and stacking fault of crack tip result in stress relief, which causes the stress to plunge. The strain of 0.08 also corresponds to the first stress peak in the stress–strain curve, which proves that titanium is sensitive to gaps, as crack serves as a determinant factor in causing the stress to plunge.

(4)   Effect of strain rate and temperature on yield point

Figure 6.17a shows the strain–stress responses of pure titanium containing an initial crack under uniaxial tension, with temperature is 300 K, crack is 10c, strain rate of 10^8 s^{-1}, 5×10^8 s^{-1}, 10^9 s^{-1} and 5×10^9 s^{-1} respectively. It can be observed from Fig. 6.17a that an important feature of the stress–strain curve is that its slope is not affected by the strain rate. When the stress reaches its peak, it plunges in these cases, while the yield stress increases with increasing strain rate, which indicates yield point is related to strain rate. This result echoes Potirniche et al. [47] who made a study on the growth and coalescence of vacancy under different strain rates.

With the increase of strain rate, the value of limiting stress is increasing. When strain rate is 5×10^9 s^{-1}, its slope is more gently than low strain rate in plastic stages. Temperature exerts great impact on material property. In the MD simulation, the temperature is set at 100 K, 300 K, 500 K and 700 K (under the transformation temperature from α-Ti to β-Ti) respectively, the strain rate is kept at 10^8 s^{-1}, and the crack is 10c. Figure 6.17b shows stress–strain curves under different temperatures. Different to the case with different strain rates, the slope of the curves is relevant to

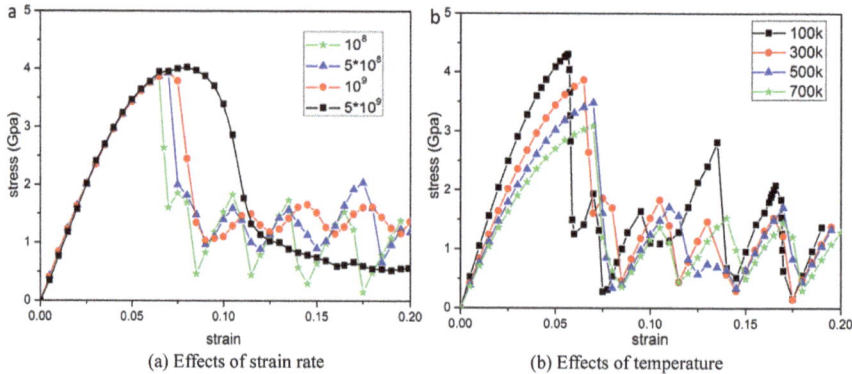

Fig. 6.17 Strain–stress response of pure titanium containing an initial crack

temperature in the stage of elastic deformation, with the increase of temperature, the slopes in elastic stage and yield point are decreasing, but the limiting strain in elastic stage is increasing. It can be observed from Fig. 6.17b that when the temperature set as 100 K, 300 K, 500 K and 700 K, the yield point is 4.304 GPa, 3.87 GPa, 3.49 GPa and 3.09 GPa respectively. It drops by about 23.33%. This is related to high-temperature degradation effect of materials. Meanwhile, after the stress reaches its peak, it plunges rapidly in each case with different temperatures.

As shown in Fig. 6.18, the stress–strain curves show wavelike shape, except the case at the strain rate of 5×10^9 s^{-1}, and with the increase of strain rate, the stress–strain curves become smooth. In order to analyze the abnormal phenomenon, Fig. 6.18 captures microstructure deformation of sample at the strain rate of 5×10^9 s^{-1}. Figure 6.18 combines microstructure and stress–strain curve, and a, b, c correspond to ε is 0.05, 0.1 and 0.15. With the increase of strain, lots of stalking faults occur, and a mass of BCC lattice are observed in the sample. However, stalking

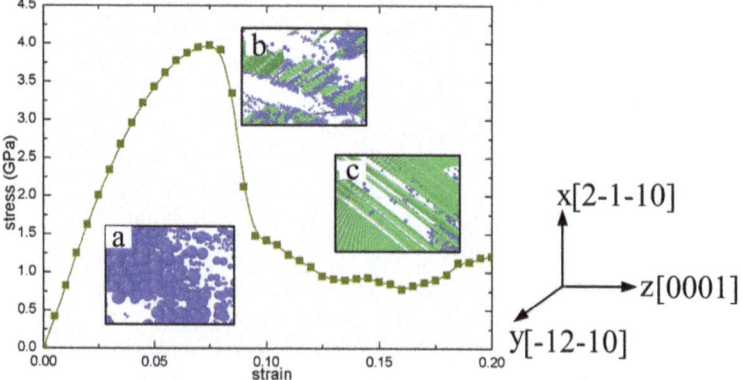

Fig. 6.18 Strain–stress curve and microstructure deformation, T = 300 K, 5×10^9 s^{-1}

faults and phase transformations are rarely observed in sample when the strain rate is 10^8 s^{-1} and 5×10^8 s^{-1}, and with the increase of strain rate, stalking faults and twinning boundary (in Fig. 6.9d, e) occur more and more, and they offer more slip plane for plastic deformation. A conclusion can be drawn from above analysis: It's stalking faults and twinning boundary that lead to the stress–strain curves become more smooth with the increase of strain rate.

In Fig. 6.18, with the variety of strain rate and temperature, the stress–strain curves show strong dependence on these variety, and the curves can all be divided into three stages: initial linear stage, sharp drop stage and wavelike decrease stage, which all coincide with Chang's study [48].

6.3.2   Molecular Dynamics Simulation of the Mechanism for Microstructure Deformation in Nano-Titanium Under Uniaxial Tension

(1)   Establishment of a molecular dynamics model of nano-titanium

For the establishment of a nano-model, an Atomsk prefabricated polycrystalline model is adopted. According to Formula (6.19), by setting the size of the model and the number of grains to change the grain size in the model, polycrystalline models with different grain sizes can be obtained.

$$\frac{4}{3}\pi R^3 = \frac{V}{n} \tag{6.19}$$

The model of the tensile deformation of nano-titanium established in this section is shown in Fig. 6.20; the model size is 20 nm $\times$ 20 nm $\times$ 20 nm, with a total of about 4.6×10^5 atoms. The lattice parameters of titanium are a = 2.95 Å and c = 4.68 Å, respectively. Red represents the titanium lattice type with a close hexagonal structure, while white represents the disordered atoms at grain boundaries. The model is tensed along the z axis. The model adopts the piston layer method (atomic layers with a thickness of about 1 nm are established at the top and bottom of the model), and uniaxial tension is simulated by tensing atomic layers. The uniaxial tension of the model adopts static-strain-rate tension, and the strain rate is set at 10^8 s^{-1}.

In this section, the Voronoi geometric method is selected to establish the initial nano model. As shown in Fig. 6.19, given the crystal grains or nucleation points during crystallization, each nucleation point expands around and grows freely, and the growth speed of each nucleation point along each direction is consistent, then the obtained grains are approximately spherical, and then the grains collide with each other to stop growing and produce grains at the junction. Satisfying the following Formula (6.20):

Fig. 6.19 Two dimensional
Voronoi structure diagram

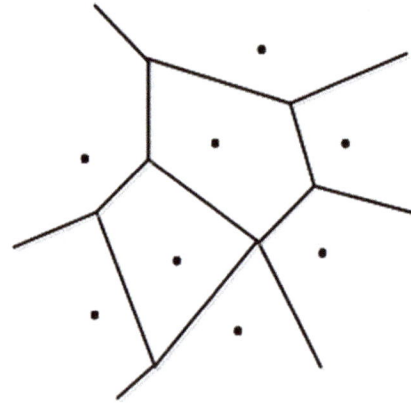

$$V_n(P_i) = \bigcap_{j \neq i}^{n} \{P | d(P, P_i) < d(P, P_j)\}(1, 2, ..., n) \qquad (6.20)$$

P^i (1, 2, …, n) is n points in two-dimensional Euclidean space, that is, the generatrix of nucleation. Vn (P^i) Voronoi polygon (i = 1, 2, …, n), where d (P_i, P_j) is the Euclidean distance.

Using the Lammps program to deal with the established model by relaxation means that the energy is minimized to maximally eliminate the forces between internal atoms. In the relaxation process, in order to eliminate the influence of boundary effect and size effect, the periodic boundary is adopted in three directions, and the conjugate gradient method provided by Lammps is adopted to obtain the nano Ti model with minimized energy. In order to further reduce the energy in the model and eliminate high internal stress still existing in the relaxation process, the Nose–Hoover method is used to control the temperature of the model, so that the system's heating-up annealing under the NPT ensemble relaxes.

Fig. 6.20 Uniaxial tension
diagram of nano titanium

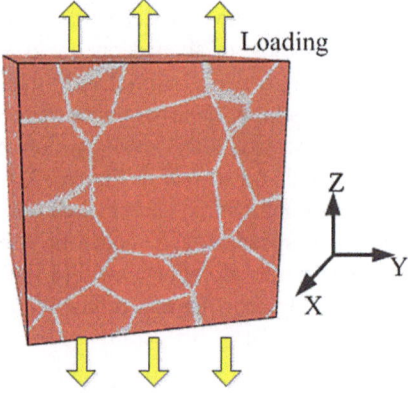

Ovito software is used for a later visualization processing, mainly using its CNA (Common Neighbor Analysis) and DXA (Dislocation Analysis) functions. Potential function is the key condition of the molecular dynamic simulation of metal materials, and it has an important influence on microstructure deformation such as atomic position, velocity and dislocation formation. Before loading the polycrystalline model, the conjugate gradient method is used to make the sample reach the state of minimum potential energy. The temperature of the external environment is taken as 300 K, the time step is taken as 0.002 ps, and relaxation is carried out for 10^5 steps.

(2)   The observation of the microstructure of crack propagation in nano titanium

Set the grain size as d = 10 nm, and the green color on the way represents FCC lattice type, and analyze the diagram of the microstructure change with a strain of 0.05, 0.1, 0.15 and 0.2 respectively. It can be seen from Fig. 6.21 that when the strain of the model reaches 0.05, stacking faults begin to appear in the model under uniaxial

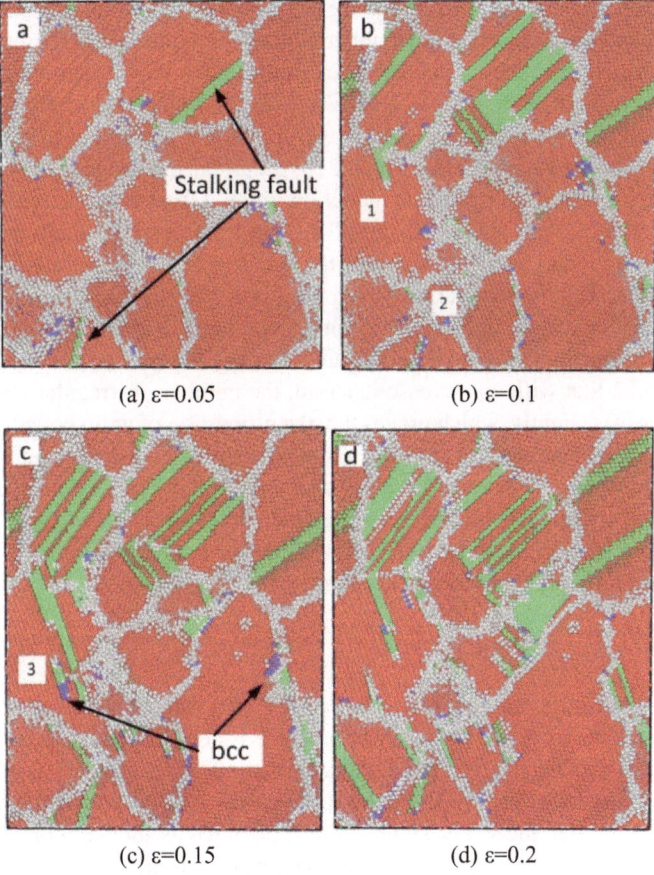

(a) ε=0.05   (b) ε=0.1

(c) ε=0.15   (d) ε=0.2

Fig. 6.21 Microstructural changes during model tensing (diameter 10 nm)

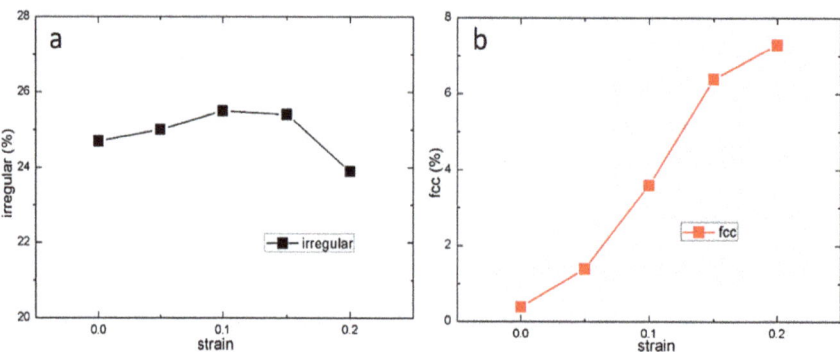

Fig. 6.22 Changes of grain boundary and FCC number during model tensing

tensile stress, and the number and density of stacking faults increase with a further increase in strain. It can be seen from Fig. 6.21b that the number of stacking faults increases with an increase in strain during the tensile deformation of nano-titanium. It is proved that the increase in the number of stacking faults is closely related to the plastic deformation of the model. Previous studies have shown that the dislocation at the stacking fault interface has a repulsive force, which effectively hinders the nucleation and slip of the dislocation, thus improving the mechanical properties of the material. It can be seen from Fig. 6.21b, c that during the tensile deformation process, the grains merge inside the model. In Fig. 6.21b, with an increase in strain, grains "1" and "2" are stacked in grain "1", and the stacking faults are emitted at the grain boundary, passing through the whole grain interior, and lead to the disappearance of grain boundaries. When the strain reaches 0.15, two grains merge into one grain. It is also confirmed that the phenomenon of grain merging and grain boundary disappearing in tensile deformation of nano titanium exists. It can be seen from Fig. 6.22 that with an increase in strain, the number of irregular atoms in the model decreases slightly, which proves that the proportion of grain boundaries in the model decreases and the number of grain boundaries decreases during the tensing process, which also illustrates the phenomenon of grain consolidation during the tensing process from the side. At the same time, the number of FCC lattice types in the model increases constantly, which proves that the transformation of lattice types occurs constantly in the process of tensing.

(3)   The effect of grain size on the mechanical properties of nano titanium

At room temperature, nano-titanium has a HCP structure. Because of its complex internal structure, there are few independent slip systems at room temperature, and it is not easy for plastic deformation to occur. Setting the grain sizes as 4, 6, 10 and 15 nm, the temperature as 300 K and the strain rate as 10^8 s^{-1}, and applying uniaxial tension to the model, the stress–strain curve is obtained as shown in Fig. 6.23:

Figure 6.23 shows the stress–strain curves with obvious differences under uniaxial tension under different grain sizes. It can be seen from Fig. 6.23 that when different

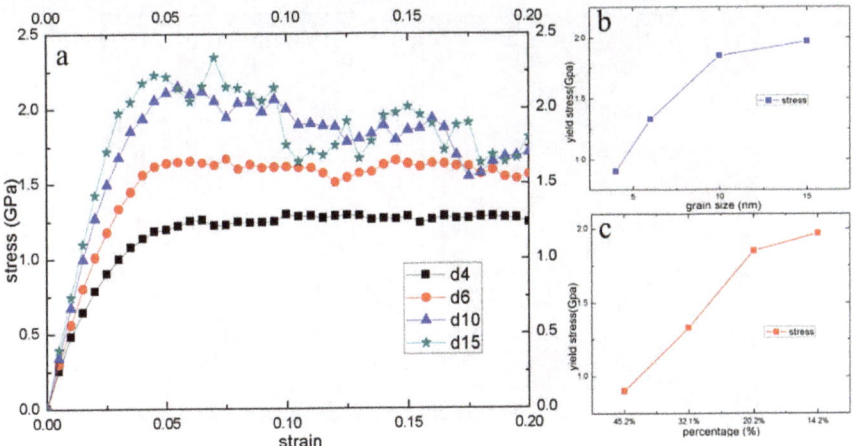

Fig. 6.23 Stress strain curves of different grain size models

grain sizes are tensed, the stress–strain curves increase to a certain peak (yield limit), and then the material enters the plastic deformation stage, and the tensile stress of the model does not drop suddenly, but fluctuates at a non-zero value, and with an increase in grain size, the yield limit of the model also increases continuously. Elastic modulus is the value of the slope of the stress–strain curve in the linear stage. It can be seen from Fig. 6.23 that the size of the grain has little effect on the elastic modulus, and with an increase in grain size, the elastic modulus of the model increases slightly. This is because with a decrease in the size of the grain, the proportion of irregular atoms at grain boundaries increases, and irregular atoms at grain boundaries are shown to be more conducive to plastic deformation, which is consistent with the previous results of FCC metals. As shown in Fig. 6.23, the abscissa indicates the size of the grain and the ordinate indicates the yield limit. It can be seen from the figure that with an increase in the size of the grain, the number of grain boundaries decreases, and the yield limit of the model increases continuously, showing the anti-Hall–Petch effect, which is consistent with the results of previous studies on the influence of the size of the grain on FCC metals (the anti-Hall–Petch effect appears when the size of the grain is less than 25 nm).

(4) The influence of the grain size on the microstructure of nano-titanium during tensile deformation

In order to further analyze the anti-Hall–Petch effect of nano-titanium, and to study the influence of the change in the grain size on the micro-deformation mechanism of nano-metal, the behavior of the tensile deformation of nano-titanium with different sizes of grain is studied at a temperature of 300 K and a strain rate of 10^8 s^{-1}.

As shown in Fig. 6.24, the diagram of the internal structure of nano-titanium with a grain size of 6 and 15 nm under strain of 0.08 and 0.16 is analyzed. It can be seen from Fig. 6.24a, b that when the grain size is small, severe deformation

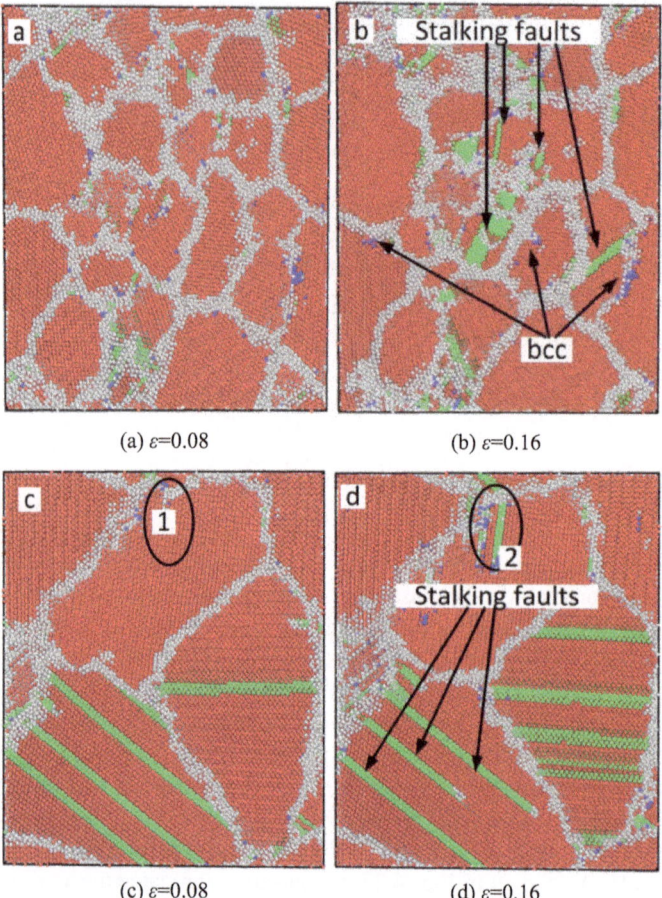

Fig. 6.24 Microstructure of different sizes of grain in the tensile process

occurs at the grain boundary and the grain shape moves or changes with the increase in strain, which proves that the movement of grain boundary is the main mode of plastic deformation of nano titanium. At the same time, a small amount of BCC phase appears at the grain boundary, and a small amount of stacking faults appear inside the grain. It can be seen from Fig. 6.24 that when the size of the grain is small, the number of stacking faults increases slowly during tensile deformation, which has little influence on plastic deformation. The plastic deformation of nano-titanium is mainly dominated by the movement and slip of the grain boundary. When the size of the grain is 15 nm, it can be seen from Fig. 6.24c, d that with an increase in the strain, a large number of stacking faults appear in the model. Comparing the areas of "1" in Fig. 6.24c and "2" in Fig. 6.24d, it can be seen that stacking faults occur at the grain boundary, start to proliferate at the grain boundary, and finally pass through the whole grain boundary. Comparing nano-titanium with a grain size of 6 nm and

15 nm, it can be seen that the plastic softening of nano-titanium is mainly caused by the movement of grain boundary when the size of the grain is small; with an increase in the size of the grain, the plastic deformation mainly depends on the slip of stacking faults.

6.4   Law of the Influence of LSP-Induced Gradient Microstructure on the Vibration Characteristics of a Thin Cantilever Beam Specimen

6.4.1   Theoretical Analysis of the Law of the Influence of Local Stiffness on the Vibration Modal Frequency of Thin Cantilever Beam Specimen

There are many thin-walled structures such as skin, blade and casing of aircraft and engine. Studies have shown that gradient structures can change the vibration charac-teristics of thin-walled structures under certain conditions. Using this characteristic of change, it can be used in the frequency modulation design of thin-walled compo-nents in the aviation field, which can effectively improve the design margin of the components.

The modal frequency of a structure is closely related to the characteristics of the stiffness of the structure. The higher the stiffness corresponding to the modal characteristics, the higher the modal frequency. In this section, based on the beam model, the mechanism of the influence of local stiffness change caused by LSP on the whole modal frequency is analyzed by means of motion analysis and energy analysis.

Establish a cantilever beam model as shown in Fig. 6.25a, with the x-axis direction as axial direction and the z-axis direction as transverse direction, and study the transverse deformation and vibration of the beam model. Here, the Bernoulli–Euler beam model is adopted, the thickness of the beam is h, the shear deformation and the moment of inertia of the section around the neutral axis are not considered, and the

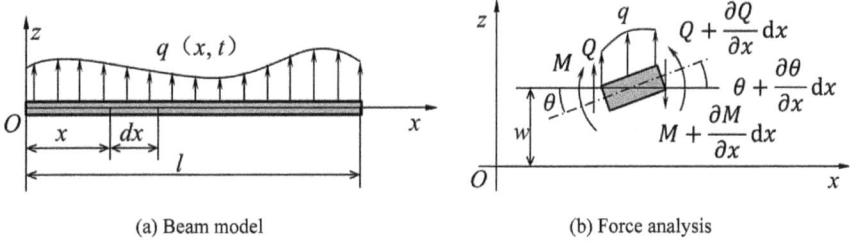

(a) Beam model   (b) Force analysis

Fig. 6.25 Beam model and infinitesimal analysis

differential infinitesimal of the beam dx, and its stress analysis is shown in Fig. 6.25b. The displacements of points on the beam in the x and z directions are recorded as u and w, respectively.

The bending stiffness of the beam model is D, which is expressed by Formula (6.21) in the direction of the thickness of the beam, regardless of whether the stiffness is distributed in gradient or uniformly, where E is the elastic modulus of the material. It can be seen that the change in the elastic modulus of the surface layer has the greatest contribution to the stiffness of the bending. Note: LSP is only performed locally in the model, so D is not uniform in the length direction of the beam, which can be expressed as a function D(x) about x.

When the beam is deformed transversely, the point strain and the normal stress on the transverse section of the beam at the distance z from the central axis are as follows:

$$D = \int_{-\frac{h}{z}}^{\frac{h}{z}} -Ez^2 dz \tag{6.21}$$

$$\begin{cases} \varepsilon_x = \dfrac{\partial u}{\partial x} = -z\dfrac{\partial^2 w}{\partial x^2} \\[2mm] \sigma_x = E\varepsilon_x = -Ez\dfrac{\partial^2 w}{\partial x^2} \end{cases} \tag{6.22}$$

According to Formula (6.22), the moment of bending produced by normal stress is:

$$M = d\frac{\partial^2 w}{\partial x^2} \tag{6.23}$$

where $\frac{\partial^2 w}{\partial x^2}$ indicates the degree of bending of the beam.

The equilibrium equation is established by the analysis of the force of the infinitesimal shown in Fig. 6.25b:

$$\begin{cases} \sum F_z = 0: \ -\dfrac{\partial Q_x}{\partial x}dx + qdx = 0 \\[2mm] \sum M = 0: \ -\dfrac{\partial M_x}{\partial x}dx + Qdx = 0 \end{cases} \tag{6.24}$$

Substitute Formula (6.23) into the second equation in Formula (6.24)-bending moment balance equation, and substitute the second equation into the first equation in Formula (6.24)-force balance equation, and eliminate dx to get:

$$\frac{\partial^2}{\partial x^2}\left(D\frac{\partial^2 w}{\partial x^2}\right) = 0 \tag{6.25}$$

When the beam model vibrates freely, q is embodied as inertial force $-\rho A \frac{\partial^2 w}{\partial t^2}$, where ρ is the density and A is the cross-sectional area of the beam, and the vibration equation can be obtained:

$$\frac{\partial^2}{\partial x^2}\left(D\frac{\partial^2 w}{\partial x^2}\right) + \rho A \frac{\partial^2 w}{\partial t^2} = 0 \tag{6.26}$$

Equation (6.26) is a fourth-order partial differential equation regarding the deflection function w. According to the specific boundary conditions and function D for the distribution of stiffness, the shape of the vibration mode of a certain mode, namely the deflection function w, and the frequency of its corresponding mode can be obtained. However, it is difficult to get general guidance directly from the equation.

The free vibration of the beam model is analyzed by the energy method, and the elastic potential energy and kinetic energy of the beam model are expressed as follows:

$$\begin{cases} U = \frac{1}{2} \int \left[D\left(\frac{\partial^2 w}{\partial x^2}\right)^2 \right] dx \\ E = \frac{1}{2} m w^2 \end{cases} \tag{6.27}$$

From the point of view of energy, the process of free vibration is also a mutual transformation process between kinetic energy and elastic potential energy. When the stiffness of the bending of the structure changes by ΔD, it can be seen more clearly from Formula (6.27) that the greater the position where ΔD acts on $\left(\frac{\partial^2 w}{\partial x^2}\right)^2$, the greater the contribution to elastic potential energy, and $\frac{\partial^2 w}{\partial x^2}$ is the degree of the bending of the beam. If the mass characteristic m is constant and the elastic potential energy increases, energy conservation can be maintained only when the moving velocity $\dot{x}$ increases, which means that the free vibration frequency increases. Therefore, it can be concluded that the greater the degree of bending deformation of the treated position, the more significantly the frequency changes.

6.4.2 Numerical Analysis of the Law of the Influence of LSP-Induced Gradient Microstructure on Vibration Modes of the Thin Cantilever Beam Specimen

This book verifies the effect of the LSPed cantilever beam specimen on modal frequency by combining finite element simulation with the modal vibration test. The structure of the designed cantilever beam with an equal cross-section is shown in Fig. 6.26, the gray part is the clamping part, and the first-order bending mode is the mode concerned by the test, which is pure bending vibration.

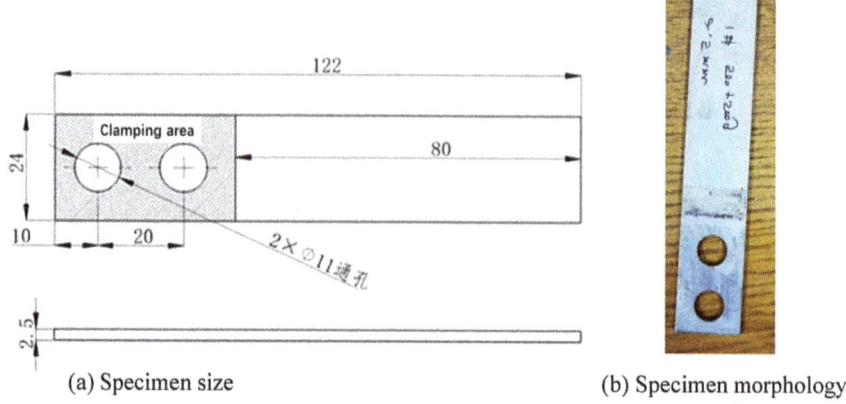

(a) Specimen size   (b) Specimen morphology

Fig. 6.26 Cantilever beam specimen

A titanium beam model as shown in Fig. 6.27 is established based on the geometric characteristics of the cantilever beam specimen, and the simulation calculation is carried out by Abaqus finite element analysis software. The model is a shell element model with geometric characteristics as shown in Fig. 6.27a, a thickness of 2 mm, constraint conditions as shown in Fig. 6.27b, model material TC-4 and parameters as shown in Table 6.3. As shown in Fig. 6.27c, the grid division adopts the quadrilateral structure grid S4R, with a grid size of 1 mm and a grid number of 2928, which meets the requirements of grid independence after analysis.

Fig. 6.27 Finite element model of a cantilever beam specimen

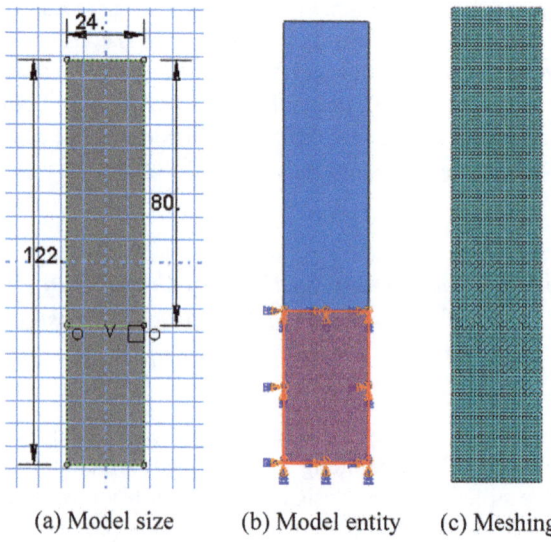

(a) Model size   (b) Model entity   (c) Meshing

Table 6.3 Main physical and mechanical properties of tc-4

Density	Elastic modulus	Poisson's ratio	Yield strength	Melting range
4.44 g/cm^3	109 GPa	0.34	827 MPa	1630–1650 °C

With the first-order mode of the titanium beam as the research object—one bending mode is taken as the research object; the modal analysis is carried out based on the Lanczos method, and the modal frequency is 318.65 Hz; the modal character-istics are shown in Fig. 6.28, and the calculated results are relative values, that is, the corresponding modal characteristics when the displacement value of the maximum displacement point of titanium beam is 1 mm. Figure 6.28a shows the characteristics of the distribution of the displacement field of each part of the titanium beam in one bending mode, that is, the deflection function in the equation in Sect. 6.3.1. It can be seen that the displacement of the clamping part and the root of the titanium beam is the smallest, and the displacement of the end part is the largest; Fig. 6.28b shows the characteristics of the distribution of the Mises stress field at various parts of the surface of the titanium beam. It can be seen that the Mises stress at the root is the

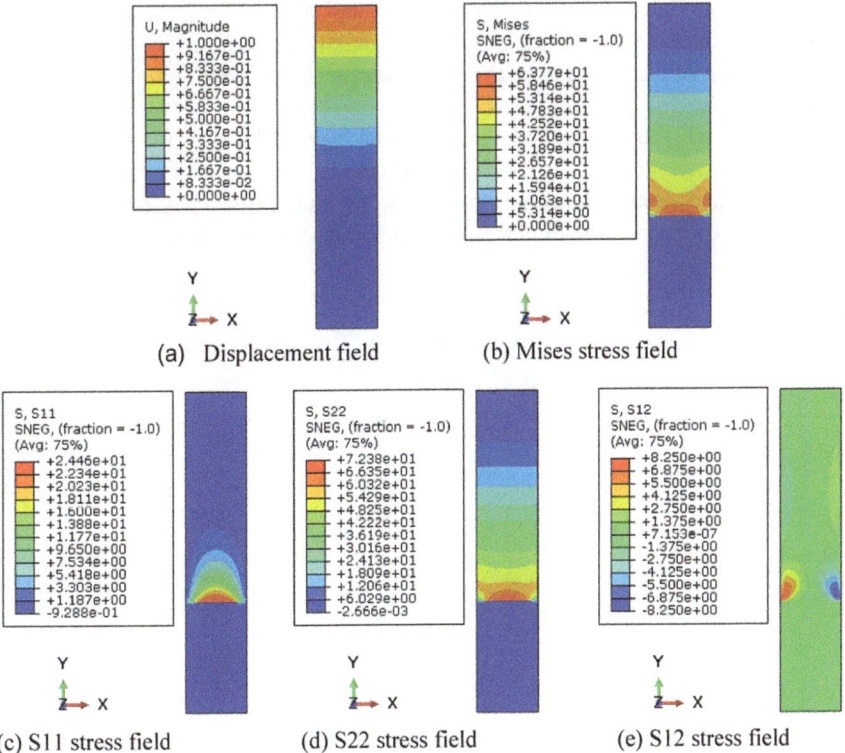

(a) Displacement field   (b) Mises stress field

(c) S11 stress field   (d) S22 stress field   (e) S12 stress field

Fig. 6.28 One bending modal characteristic of the titanium beam

largest and the Mises stress at the end is the smallest, which shows that the degree
of bending at the root of the titanium beam is the largest and the bending degree at
the end is the smallest in a bending mode. Figure 6.28c–e show the S11, S22 and
S12 characteristics of the distribution of the stress field of the titanium beam; S11
refers to normal stress in direction x, S22 refers to normal stress in direction y and
S12 refers to shear stress xy. It can be seen that, compared with S11 stress and S12
stress, S22 stress is the main component of stress, and the distribution of S22 stress
and Mises stress is close to each other, which reflects the characteristics of the strain
of the titanium beam, and the degree of bending is greater at the root and less at the
end.

Adjusting the parameters and positions of the LSP can affect the changes in the
stiffness of the components in different positions, areas and sizes, thus changing the
modal frequency.

(1)   The influence of the changes in the stiffness at different positions on modal
      frequency changes

Establish the model as shown in Fig. 6.29, in which the blue area is the clamping
area and the red area is the area of the change in stiffness. Given the original stiffness
E, the elastic modulus is set to increase by 10% and decrease by 10%, and X is given
15 values of the points in the range of [0,70], and then the influence of the change
in the position of the stiffness on the modal frequency of the titanium beam can be
obtained.

As shown in Fig. 6.30, the modal frequency and rate of change in the titanium beam
is shown when the zone of the change in stiffness is located at different positions. It
can be seen that when the stiffness increases, the modal frequency rises, and when
the stiffness decreases, the modal frequency decreases. The closer the area of the

Fig. 6.29 Position of the
change in stiffness

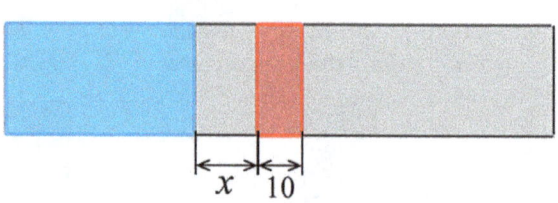

Fig. 6.30 Location of
different areas

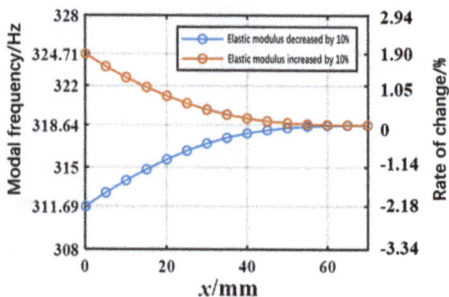

change in stiffness is to the root, the more obvious the influence of frequency is, and even the modal frequency changes by 2%. The closer to the end, the weaker the frequency influence, even close to no influence.

(2) The influence of changes in stiffness in different areas on changes in modal frequency

Establish the model as shown in Fig. 6.31, in which the blue area is the clamping area and the red area is the area of the change in stiffness. Given the original stiffness E, the rate of change in stiffness is set to increase by 10% and decrease by 10%, and X is given 15 values of the points in the range of [0, 80], then the influence of the position of the change in stiffness on the modal frequency of the titanium beam can be obtained.

   As shown in Fig. 6.32, the modal frequency and rate of change of the titanium beam is shown when the zone of the change in stiffness is located at different positions. Similarly, when the stiffness increases, the modal frequency rises, and when the stiffness decreases, the modal frequency decreases. When the area of the change of the collar is located at the root, the larger the area, there is an obvious influence of frequency.

(3) The influence of different degrees of changes in stiffness on changes in modal frequency

In the above, the stiffness is increased by 10% and decreased by 10%, respectively. In this section, the influence of the change in stiffness on the natural frequency of the structure is further analyzed, and the original elastic modulus is E = 109 GPa.

Fig. 6.31 Area of the variation of stiffness

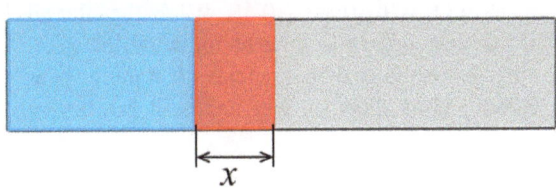

Fig. 6.32 Area of different areas

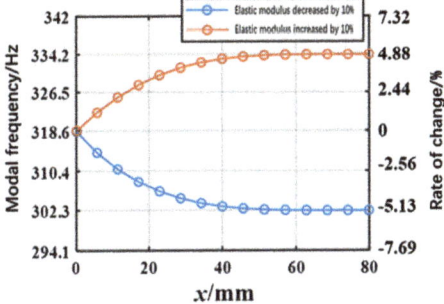

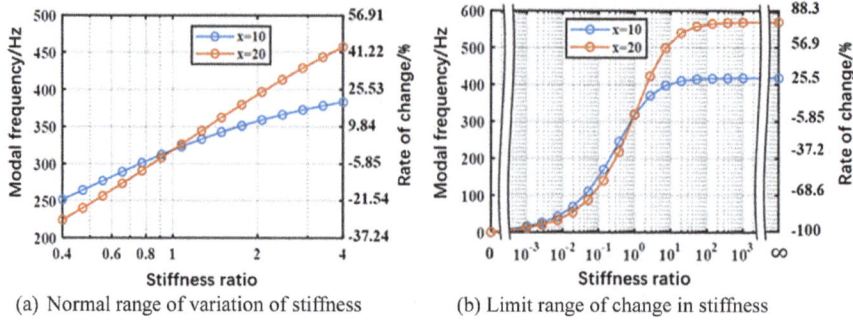

(a) Normal range of variation of stiffness   (b) Limit range of change in stiffness

Fig. 6.33 Modal characteristics of the titanium beam with a different elastic modulus

The change in stiffness of the selected area is shown in Fig. 6.32 and x = 10 and x = 20 respectively.

The local changes in stiffness are set in the conventional range and the limit range respectively, and the modal analysis is carried out, and the results are shown in Fig. 6.33a, b respectively. In the figure, the abscissa is the ratio of local stiffness to original stiffness, which is recorded as stiffness ratio, the abscissa adopts the logarithmic coordinate axis, the ordinate axis is double-axis, the left ordinate axis is the modal frequency after the changes in local stiffness, and the right ordinate axis is the rate of change in the corresponding modal frequency. As shown in Fig. 6.33b, the point where the stiffness ratio is 10^0 (i.e., point 1) is the original elastic modulus, and the corresponding modal frequency is the original frequency of 318.65 Hz.

The conventional range is determined with reference to the elastic modulus of common metals (as shown in Table 6.4). The interval of change in the elastic modulus is set as [43.6, 436], that is, [0.4E, 4E]. In the interval, 15 parameter points distributed uniformly according to logarithmic coordinates are taken for modal analysis, and the first-order natural frequencies of titanium beams with different local stiffness are obtained as shown in Fig. 6.33a. It can be seen that where the local stiffness

Table 6.4 Elastic modulus (E) and Poisson's ratio (μ) of common metal materials	Materials	E (GPa)	μ
	Tungsten	407	0.28
	Carbon steel	207	0.3
	Alloy steel	205	0.29
	Austenite stainless steel	200	0.3
	Red copper	110	0.34
	Titanium	109	0.34
	Nickel	207	0.31
	Aluminum	72	0.33
	Magnesium	45	0.29
	Brass	97	0.34

increases, modal frequency increases, and where the local stiffness decreases, and modal frequency decreases, which indicates that the modal frequency has an approximate linear relationship with the logarithm of the change in local stiffness. The trend of change in the modal frequency in different areas is the same, and where there is a larger area, there will be a greater slope of the curve.

The setting of the limit range of variation is mainly to analyze the limit of the influence of the variation of local stiffness on the natural frequency of the structure. The local elastic modulus set are first set in the interval of $\left[\frac{E}{1000}, 1000E\right]$, 15 parameter points distributed uniformly according to logarithmic coordinates are taken, and then the elastic modulus is set to 0 and $1 \times 10^9 E$ respectively, where the $1 \times 10^9 E$ is used to indicate that the elastic modulus tends towards infinity, and the modal analysis results are shown in Fig. 6.33b.

It can be seen that the trend of change of modal frequency with local stiffness is the same as that in Fig. 6.33a, but when the ratio of stiffness increases to 20, the modal frequency tends to be stable and rises slightly. When it increases to ∞, the modal frequency increases by 30.8% and 78.5% in the range of x = 10 and x = 20, respectively. It can be seen that the improvement in local stiffness has a limit to the improvement in the natural frequency of the whole structure. This is because when the stiffness is large enough, the local part can be regarded as a rigid body, and if the remaining parts of the structure can be elastically deformed, the natural frequency of the structure under this condition is the upper limit of the increase in natural frequency of the structure caused by the increase in local stiffness. When the stiffness ratio decreases to 0.02, the natural frequency changes to be stable and close to 0. When the stiffness ratio decreases to 0, the natural frequency is 0. This is because when the local stiffness of the outer part of the area of change in stiffness is reduced to 0, this part can be regarded as having no elastic restraint, so its natural frequency decreases to 0. If the local stiffness decreases to 0, the whole structure can still maintain elastic connection, and then the natural frequency of the structure is the lower limit of the reduction in natural frequency caused by the reduction in local stiffness. In short, the local part is considered as a rigid body, and the local part is not taken into consideration. Under this condition, the natural frequency of the structure is the upper and lower limits of the change in natural frequency caused by the change in the local stiffness.

6.4.3 Experimental Verification of the Law of the Influence of the LSP-Induced Gradient Microstructure on Vibration Modes of the Thin Cantilever Beam Specimen

The modal vibration test of the cantilever beam specimen is carried out, and the resonant frequency of the specimen, i.e. the natural frequency, is obtained. According to the analysis in Sect. 6.3.2, the change in stiffness at the root of the specimen

Table 6.5 Surface impact treatment scheme of plate beam specimen

	Titanium		
Treatment method	USP	LSP 1 times	LSP 3 times
Number of test pieces	8	8	8

has the most significant influence on the natural frequency of the whole structure. Therefore, set the area as shown in Fig. 6.31, set x = 10, and carry out the surface impact treatment on the area respectively. Specific implementation is shown in Table 6.5.

The test is mainly affected by processing errors and operating errors. In order to analyze the test errors and guide the design of the test scheme, eight original specimens are selected to carry out the modal vibration test, in which each specimen is tested three times, and the specimen is taken down to clean up its dust before each test, and then clamped again.

A total of 24 tests were carried out. Based on the test results, the analysis of test errors is shown in Table 6.6. In the table, Fre is the average natural frequency of the specimen, ST, STA and STE are standard deviations, indicating the error amount in Hz, where ST is the population standard deviation of test results, STA is the standard deviation between groups and STE is the standard deviation within a group. V, VA and VE are coefficients of variation, representing the error rate in %, where V is the overall coefficient of the variation of the test results, VA is the coefficient of variation between groups and VE is the coefficient of variation within a group of test results. In which ST and V represent the overall error of the tests, STA and VA represent the error between different test pieces, and STE and VE represent the error of the same test piece under different tests.

It can be seen from Table 6.6 that the intra-group error is smaller than the inter-group error, with STA reaching 7.85 Hz and VA reaching 2.75%, indicating that there is a large error between different specimens. If the control group is used for testing, the influence of local stiffness on natural frequency will be concealed by the test error. Therefore, in order to ensure the accuracy of the test, the method of intra-group comparison is adopted, and the same specimen is compared before and after treatment, that is, before LSP treatment, three resonance search tests are carried out to obtain the average value; after LSP treatment, three resonance search tests for the specimen are carried out to obtain the average value. By comparing the two values, the effect of the LSP treatment on the natural frequency is obtained.

The characteristics of the amplitude-frequency response of one specimen are shown in Fig. 6.34, and Fig. 6.34a, b show the sweep frequency results of three resonance search tests before and after LSP 1 time and LSP 3 times, respectively. It can be seen from Fig. 6.34a that the local stiffness of the titanium alloy has not

Table 6.6 Error analysis of cantilever beam specimen

Fre/Hz	ST/Hz	STA/Hz	STE/Hz	V/%	VA/%	VE/%
285.51	4.50	7.85	1.45	1.57	2.75	0.51

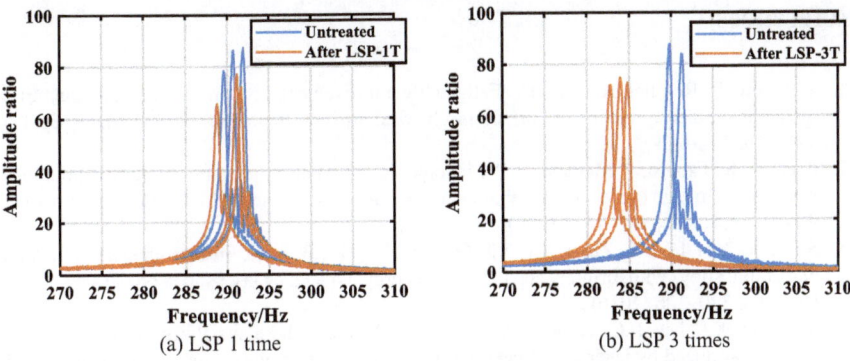

(a) LSP 1 time   (b) LSP 3 times

Fig. 6.34 Effect of LSP on the response of the titanium beam to amplitude frequency

been significantly affected by 1 time of LSP treatment. As shown in Fig. 6.34b, the resonance peak shifted significantly to the left, that is, the natural frequency decreased, which indicated that the local stiffness of the titanium alloy was reduced by 3 times of LSP, and the first-order bending mode frequency of the specimen decreased.

The overall test results are shown in Table 6.7, in which the natural frequencies before and after treatment are obtained by averaging the results of six tests of two specimens. It can be seen that the frequency of titanium beams after USP treatment increases by 0.76 Hz, with a rate of change of 0.27%, and the frequency of titanium beams after 1 time of LSP decreases by 0.32 Hz, with a rate of change of −0.11%, the absolute values of which are less than the STE and VE values of titanium beams in Table 6.6. It can be deemed that USP and 1 time of LSP have no effect on the stiffness of the titanium alloy. The frequency decreased by 7.08 Hz and the rate of change was −2.47% after 3 times of LSP. Compared with the STE and VE values of the titanium beam in Table 6.6, the change was significant, which could eliminate the influence of test error.

According to Table 6.7, the modal frequency of the titanium beam decreases by 2.47% after 3 times of LSP, and the change in the local elastic modulus after 3 times of LSP is analyzed. See Table 6.3 for setting material parameters, and the original elastic modulus E = 109 GPa. The local elastic modulus was identified by the standard LM (Levenberg–Marquardt) method combined with Abaqus modal analysis, and the local elastic modulus after 3 times of LSP was 98.07 GPa, which was 10.03% lower than the original elastic modulus.

Table 6.7 Modal frequency and change of titanium beam before and after surface impact treatment

Method	Before treatment/Hz	After treatment/Hz	Variation/Hz	Rate of change/%
USP	283.76	284.51	0.76	0.27
LSP 1 times	290.80	290.48	−0.32	−0.11
LSP 3 times	286.32	279.24	−7.08	−2.47

References

1. T. Roland, D. Retraint, K. Lu, J. Lu, Fatigue life improvement through surface nanostructuring of stainless steel by means of surface mechanical attrition treatment. Scripta Mater. **54**(11), 1949–1954 (2006)
2. P. Peyre, R. Fabbro, P. Merrien, H.P. Lieurade, Laser shock processing of aluminium alloys. Application to high cycle fatigue behavior. Mater. Sci. Eng. A-Struct. Mater. Prop. Microstruct. Process. **210**(1–2), 102–113 (1996)
3. S.H. Luo, Y.H. Li, L.C. Zhou, X.F. Nie, G.Y. He, Y.Q. Li, W.F. He, Surface nanocrystallization of metallic alloys with different stacking fault energy induced by laser shock processing. Mater. Des. **104**, 320–326 (2016)
4. S.H. Luo, X.F. Nie, L.C. Zhou, X. You, W.F. He, Y.H. Li, Thermal stability of surface nanostructure produced by laser shock peening in a Ni-based superalloy. Surf. Coat. Technol. **311**, 337–343 (2017)
5. R.Z. Valiev, R.K. Islamgaliev, I.V. Alexandrov, Bulk nanostructured materials from severe plastic deformation. Prog. Mater Sci. **45**(2), 103–189 (2000)
6. C.S. Montross, T. Wei, L. Ye, G. Clark, Y.W. Mai, Laser shock processing and its effects on microstructure and properties of metal alloys: a review. Int. J. Fatigue **24**(10), 1021–1036 (2002)
7. L.C. Zhou, Y.H. Li, W.F. He, G.Y. He, X.F. Nie, D.L. Chen, Z.L. Lai, Z.B. An, Deforming TC6 titanium alloys at ultrahigh strain rates during multiple laser shock peening. Mater. Sci. Eng. A-Struct. Mater. Prop. Microstruct. Process. **578**, 181–186 (2013)
8. A.K. Rai, R. Biswal, R.K. Gupta, R. Singh, S.K. Rai, K. Ranganathan, P. Ganesh, R. Kaul, K.S. Bindra, Study on the effect of multiple laser shock peening on residual stress and microstructural changes in modified 9Cr-1Mo (P91) steel. Surf. Coat. Technol. **358**, 125–135 (2019)
9. R.X. Zhang, X.F. Zhou, H.Y. Gao, S. Mankoci, Y. Liu, X.H. Sang, H.F. Qin, X.N. Hou, Z.C. Ren, G.L. Doll, A. Martini, Y.L. Dong, N. Sahai, C. Ye, The effects of laser shock peening on the mechanical properties and biomedical behavior of AZ31B magnesium alloy. Surf. Coat. Technol. **339**, 48–56 (2018)
10. X. Pan, S. Guo, Z. Tian, P. Liu, L. Dou, X. Wang, Z. An, L. Zhou, Fatigue performance improvement of laser shock peened hole on powder metallurgy Ni-based superalloy labyrinth disc. Surf. Coat. Technol. 126829 (2021)
11. Y. Yang, W.F. Zhou, B.Q. Chen, Z.P. Tong, L. Chen, X.D. Ren, Fatigue behaviors of foreign object damaged Ti-6Al-4V alloys under laser shock peening. Int. J. Fatigue **136** (2020)
12. M. Kahlin, H. Ansell, D. Basu, A. Kerwin, L. Newton, B. Smith, J.J. Moverare, Improved fatigue strength of additively manufactured Ti6Al4V by surface post processing. Int. J. Fatigue **134** (2020)
13. Y. Huang, S. Qu, K.C. Hwang, M. Li, H. Gao, A conventional theory of mechanism-based strain gradient plasticity. Int. J. Plast. **20**(4–5), 753–782 (2004)
14. G.I. Taylor, The mechanism of plastic deformation of crystals. Part I. Theoretical. Proc. Royal Soc. London **145**(855), 362–387 (1934)
15. G.I. Taylor, *Plastic Strain in Metals* (1938)
16. J.F. Nie, Effects of precipitate shape and orientation on dispersion strengthening in magnesium alloys. Scripta Mater. **48**(8), 1009–1015 (2003)
17. C. Domain, G. Monnet, Simulation of screw dislocation motion in iron by molecular dynamics simulations. Phys. Rev. Lett. **95**(21) (2005)
18. M. Tang, L.P. Kubin, G.R. Canova, Dislocation mobility and the mechanical response of BCC single crystals: a mesoscopic approach. Acta Mater. **46**(9), 3221–3235 (1998)
19. G.I. Taylor, The mechanism of plastic deformation of crystals. Part I. Theoretical. Proc. Royal Soc. London. Ser. A **145**(855), 362–387 (1934)
20. G.I. Taylor, Plastic strain in metals. J. Inst. Metals **62**, 307–324 (1938)
21. M.F. Ashby, The deformation of plastically non-homogeneous materials. Phil. Mag. **21**(170), 399–424 (1970)

22. J.F. Nye, Some geometrical relations in dislocated crystals. Acta Metall. **1**(2), 153–162 (1953)
23. H. Gao, Y. Huang, W.D. Nix, J.W. Hutchinson, Mechanism-based strain gradient plasticity—I. Theory. J. Mech. Phys. Solids **47**(6), 1239–1263 (1999)
24. N.A. Fleck, J.W. Hutchinson, Strain gradient plasticity. Adv. Appl. Mech. **33**, 295–361 (1997)
25. S. Qu, A conventional theory of mechanism-based strain gradient plasticity (2004)
26. J. Li, A. Soh, Modeling of the plastic deformation of nanostructured materials with grain size gradient. Int. J. Plast. **39**, 88–102 (2012)
27. N. Hfaiedh, P. Peyre, H. Song, I. Popa, V. Ji, V. Vignal, Finite element analysis of laser shock peening of 2050–T8 aluminum alloy. Int. J. Fatigue **70**, 480–489 (2015)
28. W. Braisted, R. Brockman, Finite element simulation of laser shock peening. Int. J. Fatigue **21**(7), 719–724 (1999)
29. P. Peyre, A. Sollier, I. Chaieb, L. Berthe, E. Bartnicki, C. Braham, R. Fabbro, FEM simulation of residual stresses induced by laser peening. Euro. Phys. J. Appl. Phys. **23**(2), 83–88 (2003)
30. Y. Hu, Z. Yao, J. Hu, 3-D FEM simulation of laser shock processing. Surf. Coat. Technol. **201**(3), 1426–1435 (2006)
31. J. Li, S. Chen, X. Wu, A. Soh, J. Lu, The main factor influencing the tensile properties of surface nano-crystallized graded materials. Mater. Sci. Eng. A **527**(26), 7040–7044 (2010)
32. B. Babu, L.-E. Lindgren, Dislocation density based model for plastic deformation and globularization of Ti-6Al-4V. Int. J. Plast. **50**, 94–108 (2013)
33. T.H. Fang, W.L. Li, N.R. Tao, K. Lu, Revealing extraordinary intrinsic tensile plasticity in gradient nano-grained copper. Science **331**(6024), 1587–1590 (2011)
34. Z. Cheng, H. Zhou, Q. Lu, H. Gao, L. Lu, Extra strengthening and work hardening in gradient nanotwinned metals. Science **362**(6414) (2018)
35. P.F. Wang, Z. Han, K. Lu, Enhanced tribological performance of a gradient nanostructured interstitial-free steel. Wear **402**, 100–108 (2018)
36. N.A. Fleck, G.M. Muller, M.F. Ashby, J.W. Hutchinson, Strain gradient plasticity: theory and experiment. Acta Metall. Mater. **42**(2), 475–487 (1994)
37. J.S. Stölken, A.G. Evans, A microbend test method for measuring the plasticity length scale. Acta Mater. **46**(14), 5109–5115 (1998)
38. W.D. Nix, H. Gao, Indentation size effects in crystalline materials: a law for strain gradient plasticity. J. Mech. Phys. Solids **46**(3), 411–425 (1998)
39. D. Liu, Y. He, X. Tang, H. Ding, P. Hu, P. Cao, Size effects in the torsion of microscale copper wires: experiment and analysis. Scr. Mater. **66**(6), 406–409 (2012)
40. A. Arsenlis, D.M. Parks, Crystallographic aspects of geometrically-necessary and statistically-stored dislocation density. Acta Mater. **47**(5), 1597–1611 (1999)
41. B.W. Zhang, L.C. Zhou, Y. Sun, W.F. He, Y.Z. Chen, Molecular dynamics simulation of crack growth in pure titanium under uniaxial tension. Mol. Simul. **44**(15), 1252–1260 (2018)
42. F.F. Abraham, R. Walkup, H.J. Gao, M. Duchaineau, T.D. De la Rubia, M. Seager, Simulating materials failure by using up to one billion atoms and the world's fastest computer: brittle fracture. Proc. Natl. Acad. Sci. U.S.A. **99**(9), 5777–5782 (2002)
43. Y. Mishin, M.J. Mehl, D.A. Papaconstantopoulos, A.F. Voter, J.D. Kress, Structural stability and lattice defects in copper: Ab initio, tight-binding, and embedded-atom calculations. Phys. Rev. B **63**(22) (2001)
44. D.Y. Sun, M.I. Mendelev, C.A. Becker, K. Kudin, T. Haxhimali, M. Asta, J.J. Hoyt, A. Karma, D.J. Srolovitz, Crystal-melt interfacial free energies in HCP metals: A molecular dynamics study of Mg. Phys. Rev. B **73**(2) (2006)
45. K.W. Gao, L.J. Qiao, W.Y. Chu, In situ TEM observation of crack healing in alpha-Fe. Scripta Mater. **44**(7), 1055–1059 (2001)
46. Q.P. Zhong, Z. ZH, *Fractography* (Higher Education Press, Beijing, 2006)
47. G.P. Potirniche, M.F. Horstemeyer, G.J. Wagner, P.M. Gullett, A molecular dynamics study of void growth and coalescence in single crystal nickel. Int. J. Plast. **22**(2), 257–278 (2006)
48. L. Chang, C.Y. Zhou, L.L. Wen, J. Li, X.H. He, Molecular dynamics study of strain rate effects on tensile behavior of single crystal titanium nanowire. Comput. Mater. Sci. **128**, 348–358 (2017)

Chapter 7
Study on the Compound Process of LSP and the Strengthening Mechanism on Aero-Engine Blades

7.1 Introduction

Aeroengine compressor/fan/turbine blades are the key components of turbofan engines, and their surface integrity has a major influence on the reliability and safety of the engine. However, due to the fact that the engine always works at conditions of high temperatures and high speeds, it is prone to fatigue fracture, which seriously threatens the reliability and safety of the engine. LSP is a surface treatment technology, which can effectively improve the fatigue performance of the component by forming a gradient structure with compressive residual stress and refined grains in the surface layer, and it has become a research hotspot in the anti-fatigue manufacturing field.

Surface treatment technologies, such as vibration finishing, laser additive manufacturing (LAM) and gas aluminizing, are usually used as a post-treatment process in the blade manufacturing/repair process, which is aimed at obtaining a good surface state and improving the mechanical properties. Thus, it is inevitable that LSP needs to be compounded with the existing surface treatment process of the blades, if LSP as a surface treatment process is to be applied to the manufacture/repair of aero-engine blades. However, there is little research on the compound process of LSP and the existing surface treatment technologies in the blade manufacturing/repairing process, and the effects of the compound process on the distribution of surface stress and microstructure and its strengthening mechanism are unclear.

For this reason, this section focuses on three existing surface treatment technologies of vibratory finishing, laser additive manufacturing of aero-engine compressor/fan blades and gas aluminizing of turbine blades, research on the foundation of the process and on the mechanism of the strengthening of the compound process with LSP on aero-engine blades.

© Zhejiang University Press 2021
L. Zhou and W. He, *Gradient Microstructure in Laser Shock Peened Materials*, Springer Series in Materials Science 314,
https://doi.org/10.1007/978-981-16-1747-8_7

7.2 The Compound Process of LSP and Vibratory Finishing on Titanium Alloy

Vibratory finishing is an existing surface processing technology that has been widely used in the manufacturing process of aero-engine fans/compressor blades, which is done to remove the burrs and scratches formed during the machining process of the blade, and to reduce surface roughness. Therefore, it is inevitable to determine the processing sequence between LSP and vibratory finishing if LSP is included in the manufacturing process of aero-engine blades. This section mainly studies the influence of different process treatments (vibration finishing only, LSP, LSP + vibration finishing and vibration finishing + LSP) on the quality of the surface of titanium alloy materials, and then this section determines the compound process method of LSP and vibratory finishing [1].

7.2.1 Experiments and Methods

The material of aero-engine fans/compressor blades researched in this section was TC11 titanium alloy, which is composed of hcp α phase with some residual bcc β phase. The materials were machined from a disk and were annealed at 530 °C for 6 h and air cooled, which is to relieve the residual stress to a minimum level. The chemical composition of Ti alloy is given in Table 7.1 and the material properties at room temperature are given in Table 7.2.

The samples for microstructure observations, residual stress measurements and roughness tests are schematically shown in Fig. 7.1. Four kinds of surface treatments including vibration finishing only, LSP, LSP + vibration finishing and vibration finishing + LSP, were chosen on the Ti alloy, and the sequence of the compound process was determined by residual stress and surface roughness test.

Prior to LSP treatment, the sample was mounted on a motor controlled in X–Y stage. During the LSP process, a high energy laser with a wavelength of 1064 nm and a pulse duration of 20 ns irradiated the surface of the materials, caused the

Table 7.1 Composition of Ti alloy

Composition	Al	Mo	Zr	Si	Fe	C	N	Ti
Percent (wt.%)	5.8–7.0	2.8–3.8	0.8–2.0	0.2–0.35	0.25	0.10	0.05	Bal

Table 7.2 Static properties of Ti alloy at room temperature

Yield strength $\sigma_{0.2}$/MPa	930
Ultimate tensile strength σ_b/MPa	1030
Young's modulus E/GPa	123
Poisson's ratio v	0.33

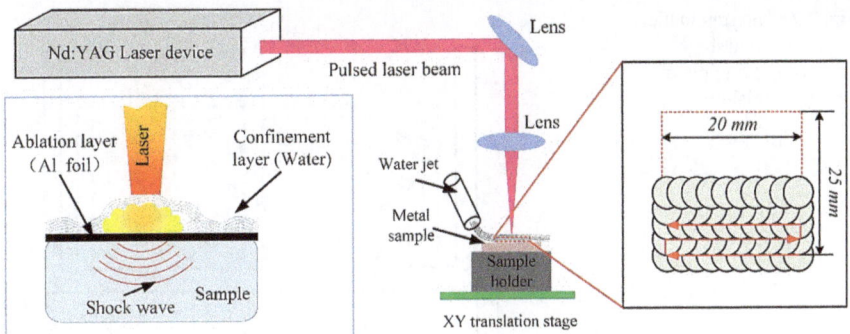

Fig. 7.1 Schematic illustrations of the LSP process. The area of samples processes by LSP for microstructure observations, residual stress and roughness tests, and the LSP paths (LSPed area with a 25 × 20 mm dimension on one side)

propagation of a high pressure (>GPa) shock wave into the material, resulting in ultra-high strain rate plastic deformation [2]. The detailed LSP process is described in Fig. 7.1. The laser energy, spot diameter, overlapping rate and repletion rate were set to 6 J, 3 mm, 50%, and 1 Hz, respectively. Thus, laser power density was about 4.24 GW/cm^2, and the corresponding peak shock wave pressure was about 3.5 GPa.

Vibration finishing refers to the relative grinding of the components to be processed by the abrasive in the tumbler at a certain vibration frequency, reducing the surface roughness of the components. During the processing, a slight plastic deformation on the surface can also be induced, thus introducing the compressive residual stress in the surface layer, and improving the surface integrity and fatigue performance [3]. The schematic illustration is shown in Fig. 7.2, the processing principle and process are briefly described as follows: First, the sample to be processed is fixed on a 5-center rotor through a fixture, and the 3-barrel is vibrated by the 4-vibration generator, and the 5-center rotor rotates around its own axis. The free rotation makes the surface of the 2-abrasive particles and the workpiece to be processed obtain a certain relative speed. Under the action of both vibration and abrasive rotation, the surface finishing of the test piece is realized. In this work, the vibratory finishing equipment was R420DL, and the abrasive materials were ceramic particles in varying shapes and sizes. The frequency was set at 50 Hz, and the duration time was 2 h.

All residual stress measurements were performed by a standard XRD technique, namely the sin2ψ-method. The pure Ti was chosen to calibrate the residual stress in accordance with the standard of ASTM E915-2010 (ASTM standard test method for verifying the alignment of X-ray diffraction instrumentation for residual stress measurement, 2010) prior to the measurement. The X-ray diffraction elastic constants were measured in accordance with ASTM E1426-2014 (ASTM standard test method for determining the X-ray elastic constants for use in the measurement of residual stress using X-ray diffraction techniques, 2014). The generator settings were 40 kV and 35 mA. The diffraction data were collected over a 2θ range from 130° to 145°, with a step width of 0.02° and a counting time of 5 s per step. To obtain the residual

Fig. 7.2 The schematic illustration and the processing principle of vibration finishing

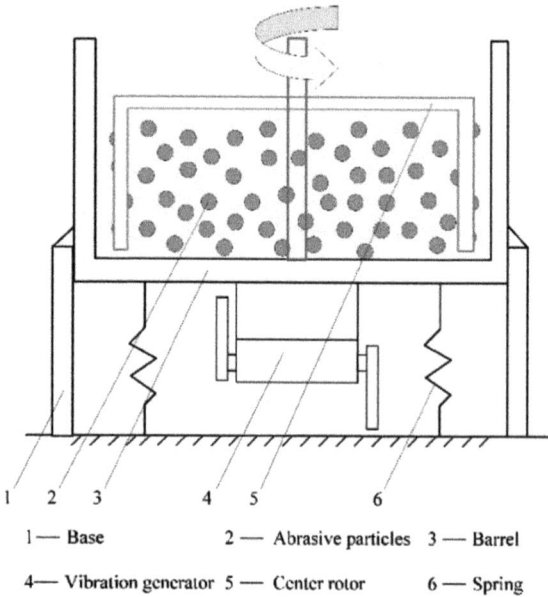

1 — Base   2 — Abrasive particles 3 — Barrel

4— Vibration generator 5 — Center rotor   6 — Spring

stress of the profiles, electro-polishing was used to obtain the different measured depths. The polishing solution was 95% saturated NH_4Cl solution $+5\%$ $C_3H_8O_3$ solution and the corrosion rate was 0.2–0.5 $\mu m/s$ each 1 cm^2. The microstructures of titanium alloy treated by LSP and vibratory finishing were studied with a TEM. TEM foils for the surface layers of the samples were prepared by mechanically grinding on the untreated sides to obtain thin plates with a thickness of ~20 μm; then the thin plates were handled by a combination of single and electro-polished twin-jets. Before putting into the TEM-3010, the foils were treated by precision ion polishing.

7.2.2   Residual Stress and Surface Roughness

The laser parameters selected is this section are a power density of 4.24 GW/cm^2 and the number of 3 impacts times according to our previous work [4]. The distributions of residual stress on the surface and profile of the specimens subjected to different surface treatments are presented in Fig. 7.3. For vibratory finishing only, the compressive residual stresses profile was rather shallow and the transition from compression to tension occurred at a depth of about 0.2 mm, and the corresponding maximum residual stress was about −100 MPa at the surface. For the LSP specimen, the largest compressive residual stress was about −760 MPa at the surface and the corresponding affected depth exceeded 1.2 mm. For the LSP + vibratory finishing specimen, the LSPed surface was continuously hit by the abrasive particles and resulted in the redistribution of the residual stress. The residual stress on the surface decreased from

Fig. 7.3 Residual stress depth profiles of Ti alloy with different surface treatments

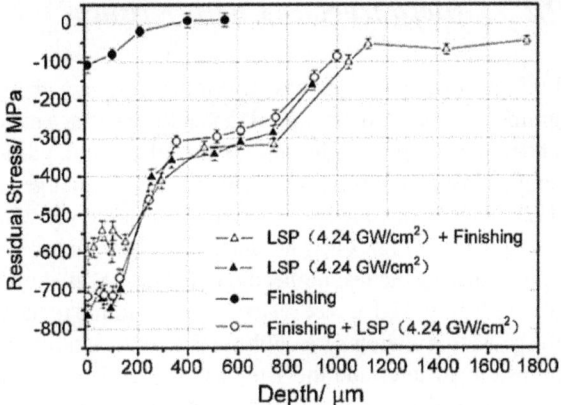

−760 to −600 MPa. The reduction of compressive residual stress may be attributed to the continual abrasion between the abrasive particles and the specimens during the vibratory finishing processing. On the other hand, because of the fact that the affected depth of the vibratory finishing treatment was only 0.2 mm, it could not affect the distribution of the pre-compressive residual stress on the sub-surface. The affected depth of compressive residual stress is determined by the propagation-affected depth of the laser-induced shock wave. This is the main reason that the affected depth of the LSP + vibratory finishing samples was consistent with that of LSP treatment (1.2 mm). For the vibratory finishing + LSP specimen, the affected depth remained unchanged, but the largest compressive residual stress was increased from −600 to −710 MPa, and the value was similar to that of LSP. The above-mentioned results showed that the sequence of LSP and vibratory finishing has no effect on the depth of the compressive residual stress of titanium alloy, but it has a certain influence on the value of the residual stress on the surface.

As we know, surface finishing has an important effect on the fatigue life of metallic alloys [4]. Therefore, to further verify the sequence of the compound process, the surface roughness after different surface treatments was measured by the TR201 hand-held surface roughness tester. The measurement values of the original and LSPed specimens were about 0.5–0.55 μm and 0.62–0.83 μm, respectively. Such values after LSP treatment are not acceptable for the application of these materials to aircraft engine blades, for which the Aeronautical Department Standard of China (HB5647-1998) set a required surface roughness of less than 0.8 μm. The surface roughness of the LSP + vibratory finishing specimen was about 0.6–0.65 μm. On the other hand, the surface roughness of vibratory finishing + LSP specimen was increased to 0.72–0.83 μm. Thus, combining the results of residual stress, the best-performing sequence of the compound process was ascertained to be LSP + vibratory finishing in this work.

7.2.3 Microstructure Characteristics

Previous research [5, 6] has shown that LSP can change the microstructure (refined grains, high dense dislocations) and induce surface nanocrystallization of the samples under appropriate conditions. On the other hand, vibratory finishing can also induce slight plastic deformation, resulting in the change in the surface microstructure. TEM images obtained from the top surface layer with different surface treatments are shown in Fig. 7.4. The untreated alloy had a dual-phased structure, with the appearance of the needle-needle like β phase, as shown in Fig. 7.4a. After the vibratory finishing, high dense dislocation was found in the top surface, which is attributed to the accumulation of a slight plastic deformation produced by the friction of the abrasion particles and the samples, as seen in Fig. 7.4b. Similarly, the grains in the near-surface region were refined into the nanoscale after LSP treatment, shown in Fig. 7.4c, d. The nanoscale grain sizes ranged between 60 and 200 nm. The corresponding circles in the selected area electrical diffraction (SAED) pattern indicated that nanocrystals were randomly oriented and many of them had high angle grain boundaries.

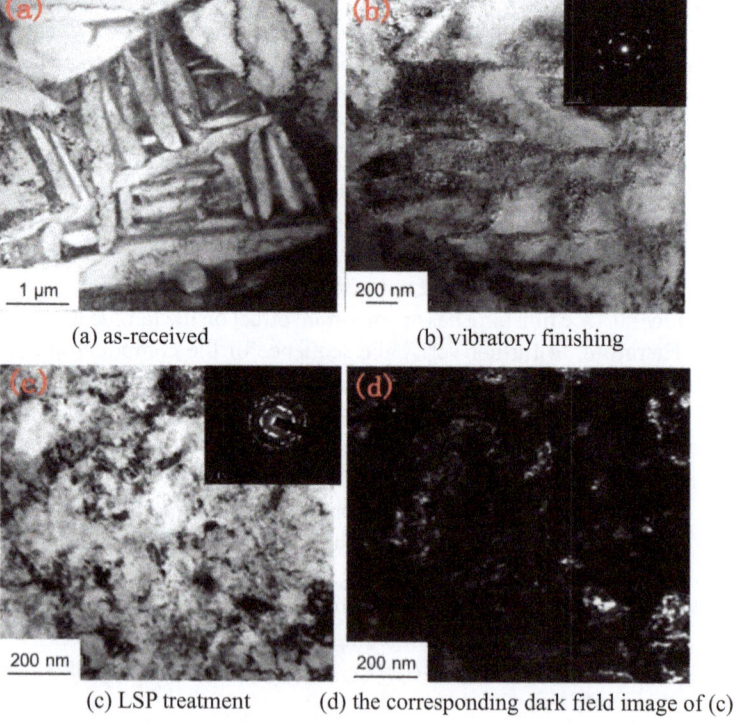

(a) as-received　　　　　　　(b) vibratory finishing

(c) LSP treatment　　　(d) the corresponding dark field image of (c)

Fig. 7.4 TEM images of Ti alloy sample

The TEM images and the corresponding SAED patterns obtained for the LSP + vibratory finishing specimen at different depths are shown in Fig. 7.5. High dense dislocations at a depth of about 20 μm (Fig. 7.5a) and both sub-structures and high-density dislocations at a depth of about 5 μm (Fig. 7.5b, c) were found. The microstructural characteristics were also verified by the corresponding SAED pattern, shown in the inset of Fig. 7.5b. The high-density dislocations and sub-structure were further changed into refined-grain and nanoscale grains on the surface, as shown in Fig. 7.5d, e. It can be found that the surface nanostructure on the surface of the specimen with compound surface treatment was more homogeneous and smaller, which shows a difference from that of the surface of the LSPed specimen. The cross-sectional microstructural characteristics were similar to those observed in the profiles of the TC6 titanium alloy [7] and TC11 titanium alloy [8] produced by LSP with the same laser parameters. Thus, we considered that the microstructural change in the sub-surface layer of the specimen treated with LSP + vibratory finishing was mainly attributed to the action of a laser-induced shock wave based on the results of Figs. 7.4 and 7.5. The pressure of the laser-induced shock wave decreased exponentially along the depth of the material. This was the main reason for the formation of gradient microstructure characteristics following the compound surface treatment.

For the LSP samples, the surface nanocrystallization of titanium alloys with high stacking fault energy was primarily attributed to the activation of dislocations [5]. According to the homogeneous nucleation theory proposed by Meyers et al. [9],

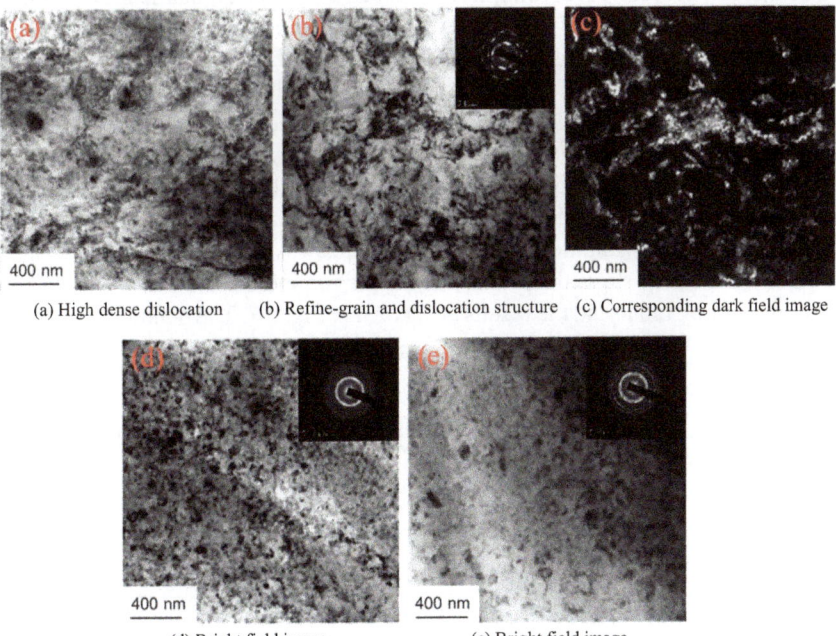

(a) High dense dislocation   (b) Refine-grain and dislocation structure   (c) Corresponding dark field image

(d) Bright field image   (e) Bright field image

Fig. 7.5 TEM micrograph of the Ti alloy treated by LSP + vibratory finishing treatment

homogeneous dislocation nucleation could be formed at the shock wave front of the material when the shock wave pressure reaches a certain threshold value (dynamic yield strength). Smaller dislocation cells were also formed under a higher shock wave pressure in an extremely short time, and finally evolved into nanoscale grains under continuous energy injection. In addition, it is likely that the plastic deformation with an ultra-high strain rate resulted in an increase in the temperature of the local material, which resulted in dynamic recrystallization [10]. In Fig. 7.5b, we have concluded that the slight plastic deformation produced by vibratory finishing can induce the formation of high-density dislocation. In other words, vibratory finishing can cause a continued movement and rearrangement of the dislocations introduced by pre-LSP treatment, which resulted in the formation of a more homogeneous nanostructure on the surface of the specimens after the compound surface treatment (Fig. 7.5d, e).

7.2.4   Fatigue Strength

Titanium alloy is the main material of aero-engine fans/compressor blades/discs, which is prone to fatigue failure under the action of high-cycle vibration fatigue load during the service process. The application of surface treatment technology is to improve the resistance to fatigue of titanium alloy. Therefore, the high-cycle vibration fatigue test was adopted for assessment and to verify the effectiveness of the compound process. According to the standard of the Aeronautical Department Standard of China (HB5277-84 Test method for vibration fatigue of engine blades and materials), eighteen fatigue samples were machined from the same process used to manufacture aero-engine compressor blades. The dimensions of the dog-bone specimens are shown in Fig. 7.6. The samples for the fatigue test were divided into two groups, one of which was treated by vibratory finishing only. The other group was treated by compound surface treatment (LSP + vibratory finishing). The LSP region and path are illustrated in Fig. 7.6.

   The finite element commercial software Abaqus was used to obtain the resonant frequency and Mises stress distribution of the fatigue specimen under the first-order vibration loading mode and the results are shown in Fig. 7.7. The boundary condition is indicated by the red dashed region fixed absolutely in Fig. 7.6. The resonant

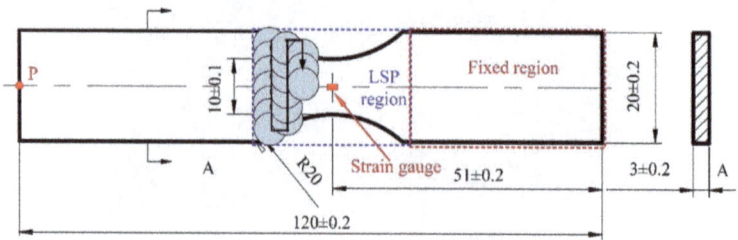

Fig. 7.6 Dimensions of the dog-bone specimens for the vibration fatigue test and LSP layout

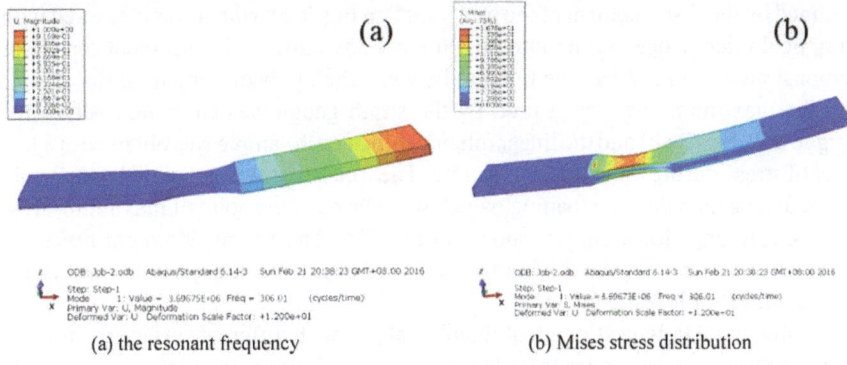

(a) the resonant frequency (b) Mises stress distribution

Fig. 7.7 The first-order mode and the Mises equivalent stress distribution of the vibration specimen

frequency of the theoretical first-order mode of the specimen was 306 Hz while deviation from the resonant frequency existed under laboratory conditions: the farther distance away from the gripping position, the greater displacement was observed (Fig. 7.7a). The Mises stress distribution of the specimen in the first-order vibration test is shown in Fig. 7.7b. The notched region of the specimen was the weakest region, where the maximum vibration stress was located. Thus, this region was confirmed to be the LSPed region, as shown in Fig. 7.6. To avoid macroscopical distortions caused by an uneven plastic distribution, both sides of the specimens were simultaneously treated with LSP with the same parameters, the detailed process was described in literature [8].

The D-300-3 electric vibration system was employed to carry out fatigue tests at resonant frequency and at room temperature, as shown in Fig. 7.8. In the system, a DYB-5 dynamic strain gauge was used as the test instrument. During the fatigue tests, specimens were fixed at one end by a clamp and free at the other end. The clamp was mounted on the vibration exciter which provided vibration displacement at the first order natural frequency of the sample. The maximum stress of specimens was

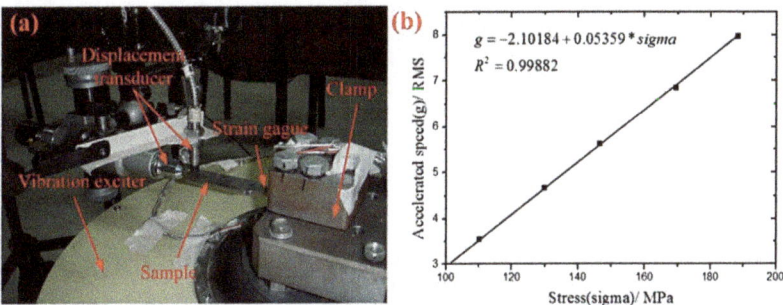

(a) Relationship between the displacement of the tip (b) The maximum stress measured by the strain gauge

Fig. 7.8 Schematic illustration of the system of the vibration fatigue experiment

obtained by the displacement of the tip (point P in Fig. 7.6), which was measured with an optical microscope and monitored with an eddy current displacement sensor, as demonstrated in Fig. 7.8a. The relation between the tip displacement of the sample and the maximum stress measured by the strain gauge was calibrated before the fatigue test (Fig. 7.8b) and the linear relationship of each sample was obtained at a low level of stress during the vibration loading. The true stress was acquired by the strain gauge during the vibrating loading, which was placed at the point of maximum stress and the reference locations are shown in Fig. 7.6. The tip displacement remained unvaried by controlling the driving frequency to ensure that the specimen works at a constant level of loading stress.

To obtain the fatigue strength of titanium alloys with different surface treatments, each group was tested under three levels of stress. Under a fixed level of stress, the test was stopped when the resonant frequency was decreased by 1% (considered as the symbol of failure of the specimens), usually prior to the complete propagation of fatigue crack through the specimen. Data from the test were plotted in the form of stress S–N curves, in which, S represents the stress that caused failure and N represents the number of cycles at which the resonant frequency decreased by more than 1%. The S–N results are shown in Fig. 7.9, from which it can be clearly seen that after the compound surface treatment, the fatigue limit improved from 438 to 544 MPa, a 24.2% increase. The results indicated that compound surface treatment could effectively improve the fatigue strength of Ti alloy.

The JEOL/JSM-6360LV scanning electron microscope was used to observe the fracture morphologies of the vibration specimens after the fatigue test. The comparison of the crack initiation sites and propagation zones with different surface treatments is to analyze the mechanism for the effect of the compound process on the fatigue strength, the results are shown in Fig. 7.9. It can be seen that the crack initiates on the surface due to the thin layer affected by the vibratory finishing, which is barely influenced on the initiation of fatigue cracking, as seen in Fig. 7.10a. The previous result shows that the depth of the compressive residual stress is larger than

Fig. 7.9 Stress against number of cycles (S–N curves) of the Ti alloy with vibratory finishing and LSP treatment

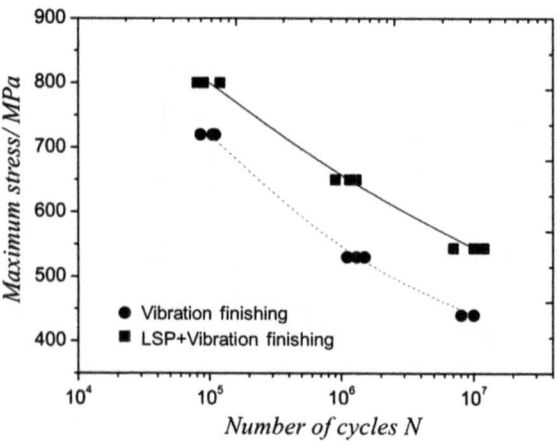

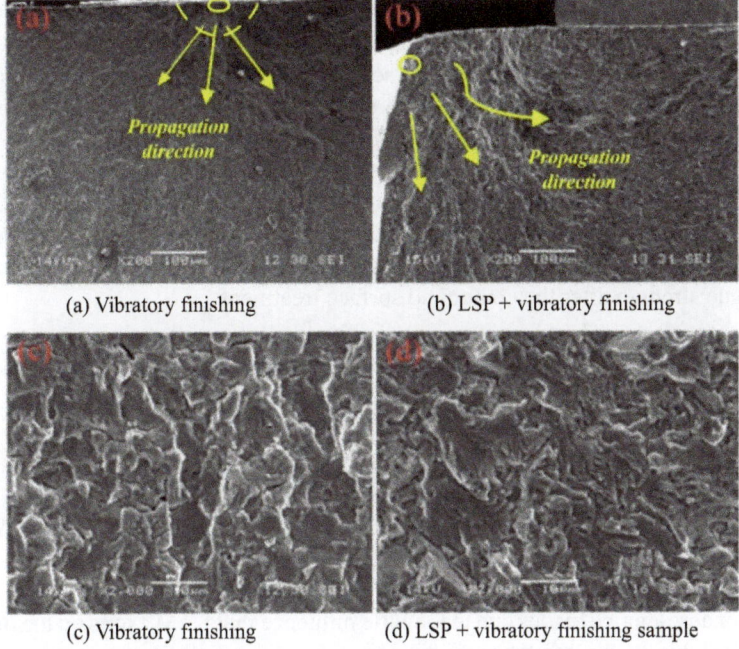

      (a) Vibratory finishing   (b) LSP + vibratory finishing

      (c) Vibratory finishing   (d) LSP + vibratory finishing sample

Fig. 7.10   Typical fatigue fractography of specimens after fatigue test

1.5 mm after one-sided LSP (Fig. 7.3), and the stress can still exceed the dynamic yield strength at the mid-plane (1.5 mm depth) although the stress wave declined. Thus, plastic deformation could be produced and the compressive residual stress was induced at the mid-plane. According to the results from Zhang et al. [11], when the shock wave pressure increases beyond 3 GPa, the corresponding residual stress on the cross-section of the titanium alloy with a thickness greater than 3 mm presents a compressive–tensile–compressive–tensile–compressive profile after a two-sided LSP treatment. In this section, the shock wave pressure is 3.5 GPa, a similar distribution of residual stress in depth can be induced. Due to the existence of compressive residual stress on the surface layer, the crack initiation site moved into the sub-surface with a distance of about 50 μm from the surface after the two-sided LSP treatment. In parallel, there were some small regions of smooth surfaces formed near the initiation of fatigue cracking and the path of the fatigue crack was more tortuous, as shown in Fig. 7.10b. The more grain boundaries and complicated distribution of residual stress after the compound surface treatment, more cleavage steps were generated in the flat region, as shown in Fig. 7.10c, d.

As we known, the location of the initiation of fatigue cracking was directly relevant to the complex microstructural characteristic and stress distributions that were formed. The surface nanostructure and high-density dislocations in the sub-surface were found after compound surface treatment (Fig. 7.5). More grain boundaries were found in the refined grains on the surface, which can offer a higher energy barrier, restraining the generation of slip bands and the growth of cracks. The high-density dislocation in the subsurface layer can also effectively restrain the plastic flow of the material and hinder the propagation of cracks. And the energy stored in the dislocation structure during plastic deformation can effectively reduce the driving force of the propagation of cracks during fatigue loading. In addition, the considerable compressive residual stress presented in the specimens after the compound surface treatment could decrease the mean stress applied and the stress intensity factor according to the Goodman theory [12], and effectively delay the initiation of fatigue cracking and reduce the rate of the growth of cracks, as demonstrated by the remarkable increase in fatigue strength after the compound surface treatment.

7.3 Regain the Fatigue Strength of LAMed Titanium Alloy via LSP

Laser additive manufacturing (LAM) is a kind of technology for rapid repair which can rebuild the complex geometrical shape of damaged components and restore some mechanical properties. Thus, this technology has been applied in the manufacture and repair of aero-engine blades due to the little influence that LAM exerts on the matrix. However, due to the existence of a large thermal gradient in the repair process, residual tensile stress, relatively loose microstructure and even micro-cracks are formed, affecting the fatigue performance of the repaired parts. Therefore, reducing or eliminating the residual tensile stress and improving the microstructure are the key problems to be solved in the application of LAM technology in the repair of damaged titanium alloys. LSP is a kind of technology that modifies the surface, which can optimize the distribution of the field of stress on the surface of the material and change the microstructure of the surface. Therefore, LSP can be used as a post-treatment technique in repairing damaged blades via LAM [13].

7.3.1 Experiments and Methods

The material investigated in this section was TC17 titanium alloy, which has been widely used for materials for fans or compressor blades in advanced aero-engines, with a typical microstructure of bimodal distribution of α phase and β phase. Samples with the dimension of 170 mm × 200 mm × 6 mm were supplied after having been annealed at 800 °C for 4 h and followed by water quenching; the chemical composition is shown in Table 7.3.

The LAM experiment was conducted on an independently-developed system equipped and designed by Northwestern Polytechnical University with a Rofin Sinar 850 CO_2 laser, the laser parameters are listed in Table 7.4 and Argon was used as a

Table 7.3 Chemical composition of TC17 titanium alloy

Composition	Al	Mo	Cr	Sn	Zr	Fe	O	C	Ti
Percent (wt%)	4.5–5.5	3.5–4.5	3.5–4.5	1.6–2.4	1.6–2.4	0.30	0.13	0.05	Balance

Table 7.4 Parameters employed in the laser additive manufacturing process

Parameter	Value	Parameter	Value
Laser power (W)	2000	Scanning speed (mm/s)	10
Laser spot (mm)	2	Powder flow rate (g/min)	4
Laser overlapping ratio	25%	Gas flow rate (L/min)	5
Powder material	TC17 titanium	powder diameter (μm)	75
Type of gas	Ar		

protective gas during the LAM process. The process of the preparation of the specimen is shown in Fig. 7.11. A representative in-service grind-out was applied on each plate using a wire-electrode cutting and was filled by LAM. The TC17 titanium powders with particle diameters of 75 μm, protected in a highly pure argon atmosphere, were pre-placed on the substrate with a thickness of approximately 3 mm

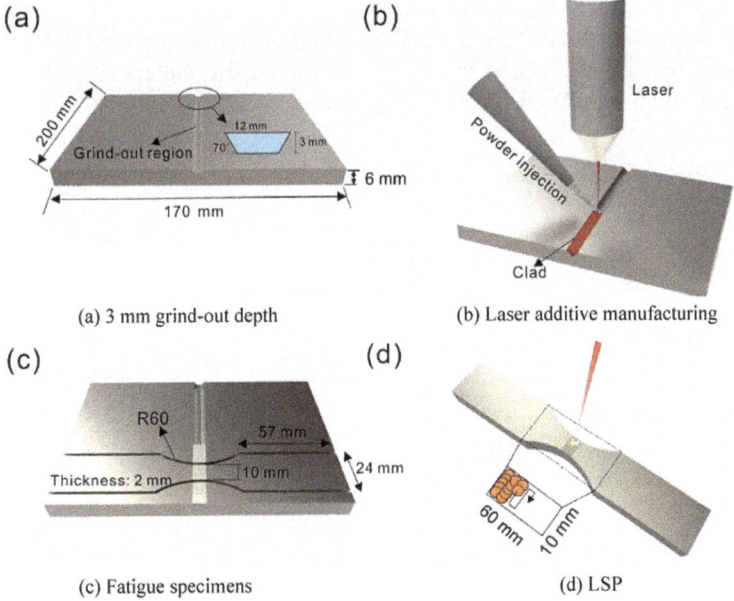

(a) 3 mm grind-out depth

(b) Laser additive manufacturing

(c) Fatigue specimens

(d) LSP

Fig. 7.11 The schematic drawing of laser processing and the preparation of the fatigue specimen

(Fig. 7.11a). Between each two successive layers, the sample was moved by 0.3 mm vertically, defining the height of each layer (Fig. 7.11b). Due to the thick of the original plate with 6 mm, there no significant distortion was produced after laser additive manufactured. Then the laser additive manufactured plate was removed by the machining to obtain a slab with the thickness more than 2 mm, and subsequently the excess layer was progressively removed by SiC papers with grit number from 800 to 2000 to gain a completely flat plate with 2 mm (Fig. 7.11c), the surface roughness Ra was 0.419 μm, as seen in Fig. 7.12a. The laser additive plate was wire-cut into dog-bone fatigue specimens with the final dimensions according to the Aeronautical Department Standard of China (HB5277-84), as seen in Fig. 7.11c. Lastly, LSP treatment was performed on the fatigue specimens with the Ra $\leq$ 0.32 μm, as shown in Fig. 7.11d.

The specimens with a rectangle area of 60 mm $\times$ 10 mm were treated with a low-power density by double-sided LSP (Fig. 7.11d), which was to avoid macro deformation due to the merely 2 mm thickness of the specimen. During the LSP processing, an Nd:YAG laser with a wavelength of 1064 nm and a pulse of around 20 ns was conducted, a water layer of about 1 mm in thickness was used as the transparent confining layer and an Al foil with a thickness of 100 μm was used as the absorbing layer. The laser parameters were as follows: a laser energy with 2.8 J, spot size with the diameter of 2.4 mm, the spot overlap of 50% and the three impacts. After LSP treatment, the roughness of the surface was increased to 0.704 μm, as seen in Fig. 7.12b. The results showed that LSP has little influence on the roughness of the surface, which is suitable for blade processing.

A Proto-LXRD instrument was used to measure residual stresses of the LAMed Ti-alloy with and without LSP treatment in direction X (Fig. 7.13c) using conventional X-ray diffraction via the sin2φ method. Details of the parameters for the measurement of residual stress are provided in Table 7.5. To compare the extent and depth of plastic deformation, \FWHM values were also recorded through the depth for the

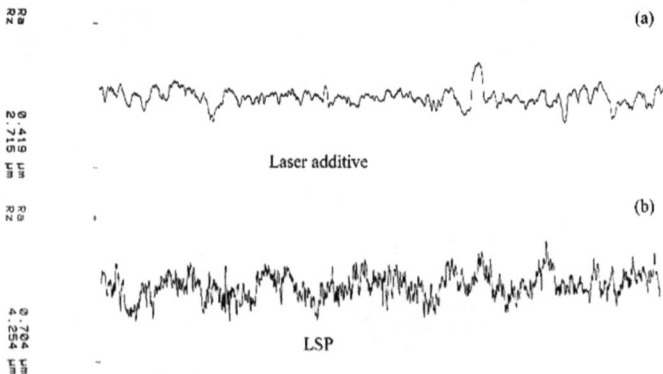

Fig. 7.12 The roughness of the surface of the specimen with a different surface treatment. **a** Laser additive manufactured Ti-alloy with surface polish; **b** LSP treatment on the laser additive manufactured Ti-alloy

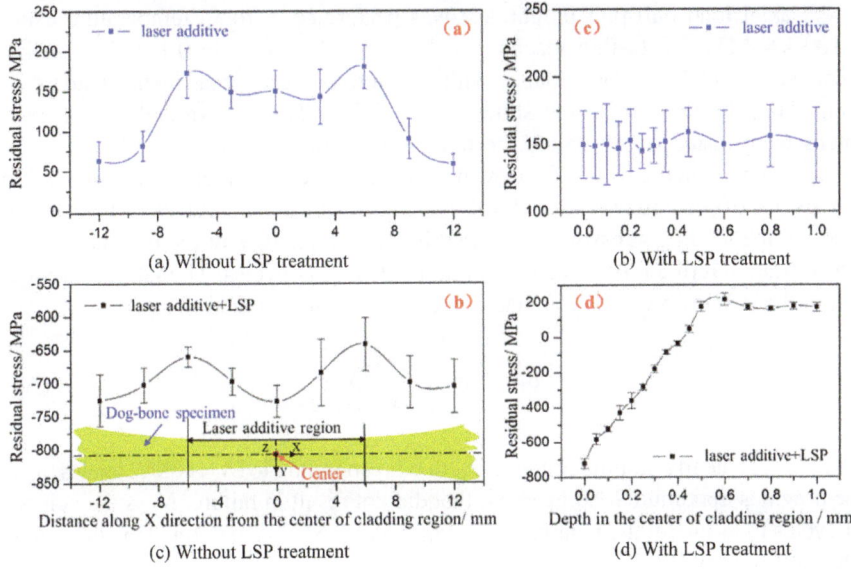

Fig. 7.13 Distribution of residual stress with and without LSP treatment in the laser additive region and the xyz direction of the measurement of residual stress on the specimen was seen in (b)

Item	Description
Detector	PSSD (Position sensitive scintillation detector), 15° 2θ range
Radiation	Cu-kα (λ = 1.540562 nm)
Tilt angles	0°, ±4.64°, ±19°, ±25°
Aperture size	3 × 1 mm rectangle
Plane; Bragg's angle	Set of {213} planes; 142°

Table 7.5 Parameters for XRD residual stress measurement

same peak with and without LSP treatment in the laser additive region. The relevant test standards refer to Sect. 7.2.1.

Surface and cross-sectional microstructural characteristics of laser additive manufactured specimens with and without LSP treatment were observed via a JEOL/JSM-6360LV SEM operated at a voltage of 15 kV. The samples to be observed were corroded with a solution made up of 92% distilled water +3% HF +5% HNO$_3$ (volume ratio). A TEM-3010 TEM was adopted to precisely characterize the surface microstructure of the laser additive specimens before and after LSP treatment. The thin foils for TEM observations were cut by electro-spark discharge from the surface layer with a thickness of 1 mm; then they were mechanically polished on the untreated sides to obtain thin plates with a thickness of about 20 μm, and lastly, the thin plates were prepared by a combination of single and twin-jet electro-polished to make it suitable for TEM observation.

An axial load pull-pull fatigue test was performed at room temperature using a 100 kN MTS810 testing machine with a stress ration R = 0.1 and a loading frequency of 15 Hz, in accordance with the Aeronautical Department Standard of China HB5287-96 (dimensions shown in Fig. 7.11c). For each variable, three specimens were tested to determine the fatigue strength at the life of ΔN cycles by a step-loading method [14]. In this method, a single specimen was subjected to ΔN cycles of loading at a stress level (σ_0). If the specimen survived at this step of loading, the loading stress was increased by $\Delta\sigma$ (about 10% of the σ_0) and the test was repeated until the specimen failed in less than ΔN cycles. The fatigue strength σ_{FS} of the specimen can be calculated in the following equation:

$$\sigma_{FS} = \sigma_0 + \frac{N_f}{\Delta N}\Delta\sigma \qquad (7.1)$$

where σ_{FS} is the maximum fatigue strength corresponding to $(N_f + \Delta N)$ cycles, σ_0 is the previous maximum loading stress that did not result in failure, N_f is the number of cycles to failure at the loading stress ($\sigma_0 + \Delta\sigma$), ΔN is the defined cyclic fatigue life (in this work it is 2×10^5).

7.3.2 Distribution of Residual Stress

The results of the measurement of residual stresses in the central additive region and the HAZ without and with LSP treatment are shown in Fig. 7.13. It can be clearly seen that after LAM repair, a large amount of residual tensile stress is generated on the surface of the repair zone and the heat-affected zone, among which the value of the residual tensile stress on the surface of the repair zone is about 150 MPa, and the value of the residual tensile stress in the heat-affected zone increases to 180 MPa. The residual stress in the direction of the depth of the central position of the repair area (marked position in Fig. 7.13c) was further tested, the tensile stress in the whole area was between 140 and 160 MPa. The reasons for the introduction of residual tensile stress were mainly including two stages [15]. First, the laser-induced heated region expanded against the restriction imposed by the unaffected cold substrate during the LAM process. Second, volumetric shrinkage occurred to the added material and the HAZ region due to the severe temperature drop, and then the laser additive was again restrained by the larger elastic region of the substrate material. In other words, the residual tensile stress was a result of the thermal gradient during the laser additive process where the colder material constrained the hotter material. Meanwhile, the large volumetric expansion and shrinkage in the HAZ region could be produced during the whole process, which resulted in the formation of the largest tensile stress with about 180 MPa.

After LSP treatment, there was a significant change in the distribution of surface residual stress, and the residual stress changed from tension to compression in the

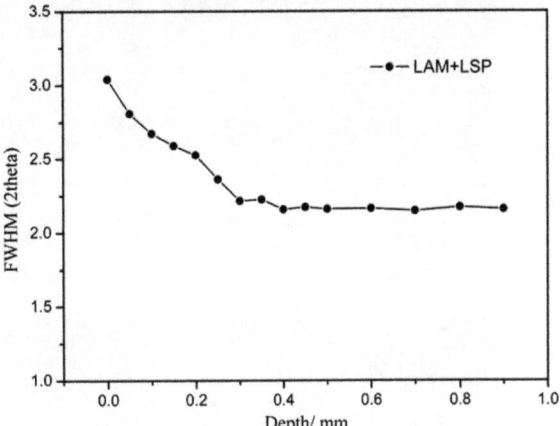

Fig. 7.14 FWHM versus depth in the center of the laser additive specimen with LSP treatment

two regions. The maximum residual compressive stress in the center was −727 MPa (Fig. 7.13b). Furthermore, the cross-sectional stress in the center of laser additive region was measured. Due to the specimens with double-side LSP, it would be thought that the distribution of residual stress in depth was symmetrical regarding the middle-line cross-section. The tensile residual stresses were turned into compressive residual stress and the values decreased gradually along the depth from the surface and remained in the compression region up to a depth of at least 0.5 mm (Fig. 7.13d), which was consistent with the attenuation of the pressure of the shock wave in the material.

XRD peak broadening is characterized by FWHM values, which was a quantitive indication of the level of plastic strain introduced in the laser additive specimen, were recorded during the measurements of residual stress, as shown in Fig. 7.14. The level of plastic strain was higher at the surface (FWHM of ~3.04) and subsequently decreased through depth to a more or less constant value after 400 μm.

7.3.3   Characterization of the Microstructure

Figure 7.15 showed the typical surface microstructure in three different regions, namely the substrate region, the laser additive region and the HAZ region, of the laser additive specimen. Figure 7.15a shows the SEM morphology in the top surface of the substrate region and Fig. 7.15d is the magnified image of the quadrangle in Fig. 7.15a. The typical α + β structure, coarse parallel β phases with a large number of acicular structures and a small amount of lath structures, coarse columnar α phases were clearly found. In addition, a bit of α phase with several micrometers exist in the acicular structure as found (Fig. 7.15d). Figure 7.15b shows the SEM morphology in the top surface of the laser additive region and Fig. 7.15e is the magnified image of the quadrangle in Fig. 7.15b. Due to the high temperature gradient and the rapid rate of

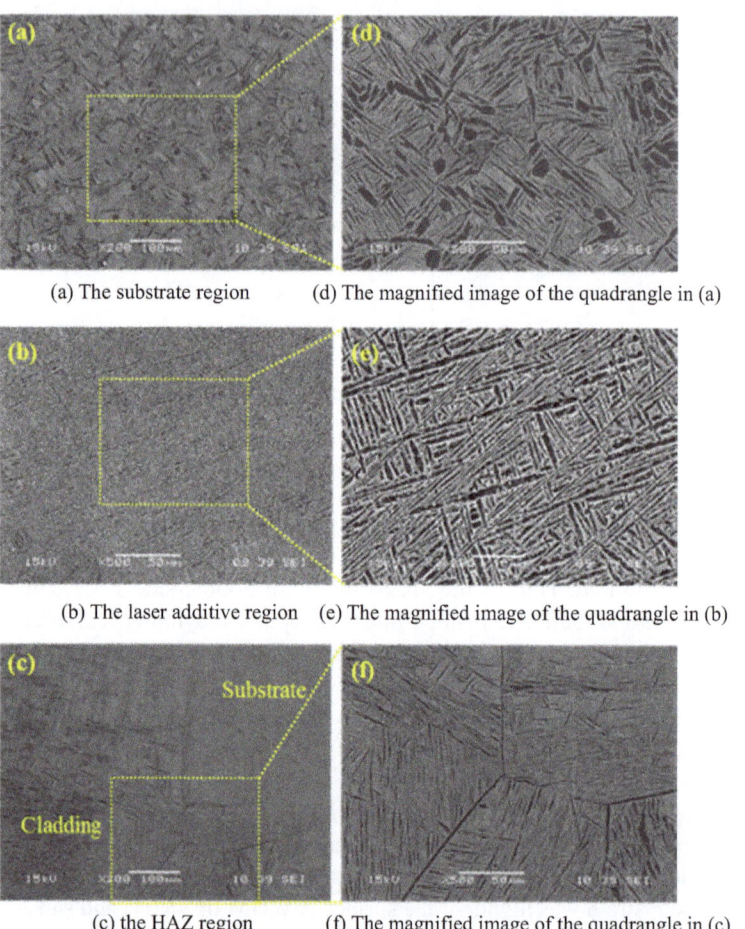

(a) The substrate region   (d) The magnified image of the quadrangle in (a)

(b) The laser additive region   (e) The magnified image of the quadrangle in (b)

(c) the HAZ region   (f) The magnified image of the quadrangle in (c)

Fig. 7.15 The surface microstructure in different regions of the laser additive specimen

cooling in the LAM process, a superfine basket-weave structure within the columnar grains was formed in the repair region (Fig. 7.15e). This is why we chose a larger magnification in this region compared with that in the substrate and transition regions. Due to the characteristics of the preferential growth of the titanium alloy and the large temperature gradient, a small number of columnar crystals gradually expanded outward during the LAM processing. The distance between columnar crystals was less than 1 μm, which is less than the original structure of columnar crystals, and the α phase also had a similar phenomenon. Figure 7.15c shows the cross-sectional SEM morphology in the top surface of the transition region and Fig. 7.15f is the corresponding magnified image. An obvious difference in microstructure occurred, and the basket-like structure disappeared and was replaced by the biphasic structure. Due to the higher diffusion coefficient of the β phase atoms in titanium alloy, the size of the acicular β phases became coarse grains. The detailed discussion about the effects of LAM on the surface microstructure of titanium alloys was reported by Zhu et al. [16].

Figure 7.16 shows the typical cross-sectional microstructure in three regions of the LAM + LSP specimen. Figure 7.16a presents the cross-sectional SEM morphology of the substrate region, and Fig. 7.16d is the magnified image of the quadrangle in Fig. 7.16a. The grain sizes were refined compared with those of the specimen without LSP treatment (Fig. 7.15d) in the same region, and the grain sizes present gradient distribution, which was attributed to the strain rate imported by LSP decreased along the depth from maximum (up to 10^7 s^{-1}) at the top surface to zero. The similar

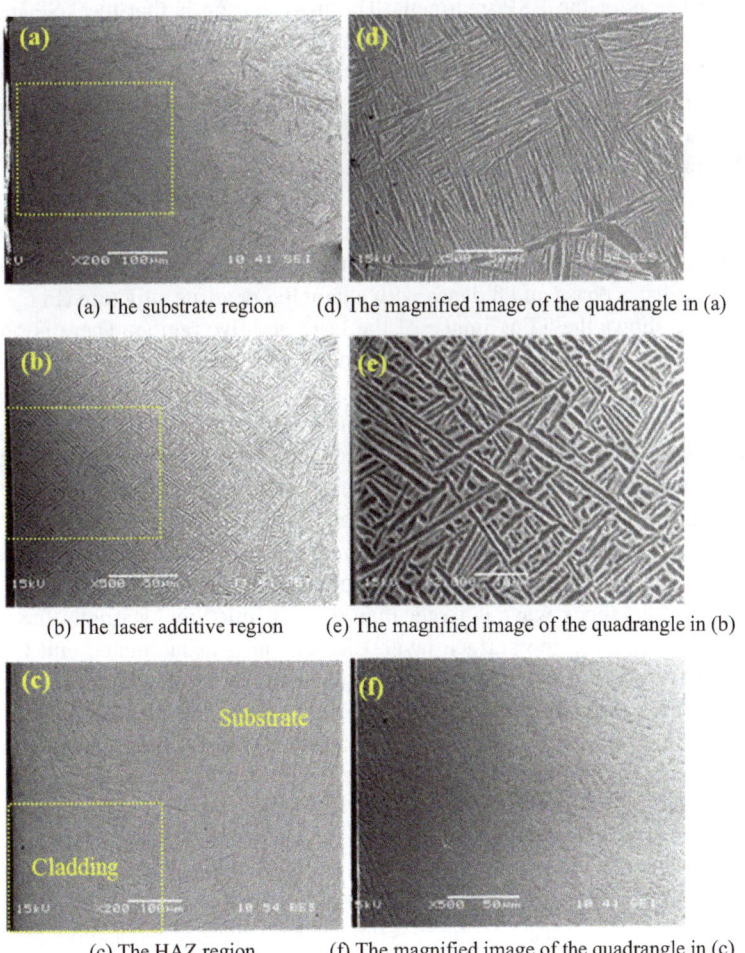

(a) The substrate region   (d) The magnified image of the quadrangle in (a)

(b) The laser additive region   (e) The magnified image of the quadrangle in (b)

(c) The HAZ region   (f) The magnified image of the quadrangle in (c)

Fig. 7.16 Typical cross-sectional microstructure in three regions of the LSP post-laser additive specimen

results have been reported in literatures [5, 6]. Compared with the structure of the laser additive region in Fig. 7.15e, there was no significant change in the legible basket-weave structure with columnar α and acicular β structure, shown in Fig. 7.16e (the magnified image of the quadrangle in Fig. 7.16b) after LSP treatment, and the distance between columnar crystals increased compared to that of the LAMed specimen. Figure 7.16c shows the cross-sectional SEM morphology of the HAZ region, and Fig. 7.16f is the magnified image of the quadrangle in Fig. 7.16c. The microstructure is different from that in this region of the laser additive specimen, no obvious phase boundary was found.

According to the results of the SEM observation, LAM repair could effectively change the microstructure of the material, and there was no obvious change in the microstructure after the LSP treatment. The main reason is that the LSP-induced plasma shock wave acts on the surface of the material, causing the material to undergo plastic deformation of an ultra-high strain rate. Meanwhile, the cold hardening rate of the material is low, and the layer of plastic deformation formed in the surface layer is small.

TEM observations were adopted to further analyze the effect of LSP on the surface microstructure of the different region of the LAMed Ti alloy specimen. Figure 7.17a–c shows the TEM images of the top surface in different regions of the LAMed specimen without LSP treatment. The original Ti alloy is made up of two phases with large sizes: α phase and β phase with a lamellar dendrite structure (Fig. 7.17a). Figure 7.17c shows the TEM image of the laser additive region. There is obvious difference with that in the substrate region, and it can be seen that typical characteristic with a lot of acicular and lathy β-phases precipitated from the nascent columnar crystals, and the sizes of the dual-phase became smaller than what they were in the substrate region. Figure 7.17b shows the microstructure features in the HAZ region. The microstructure is similar to that in the laser additive region, and the sizes were larger. Figure 7.17d–f shows the TEM images in different regions of the LAMed specimen after LSP treatment. For the substrate region, refined grains are observed (Fig. 7.17d), which is attributed to the dislocation movements. When the pressure of the laser-induced shock wave is larger than the dynamic yield strength, the plastic deformation occurs on the surface layer, dislocation is accumulated and then the sub-boundary forms, lastly results in the refinement of coarse grain. The detailed grain refinement mechanism has been described in detail in a previous study [5]. In the laser additive region, a large amount of high dense dislocations in the grain can be found, as shown in Fig. 7.17f. Due to the faster cooling rate in the HAZ region than that in the laser additive region and the relatively lower temperature, the porous microstructures in this region were generated, which resulted in a relative large space between the adjacent grains [17]. Therefore, the laser-induced shock wave decayed rapidly in this region and resulted in a smaller plastic deformation compared with that in the laser additive region. This was why no high-density dislocations, and only dislocation lines were formed in this region (Fig. 7.17e).

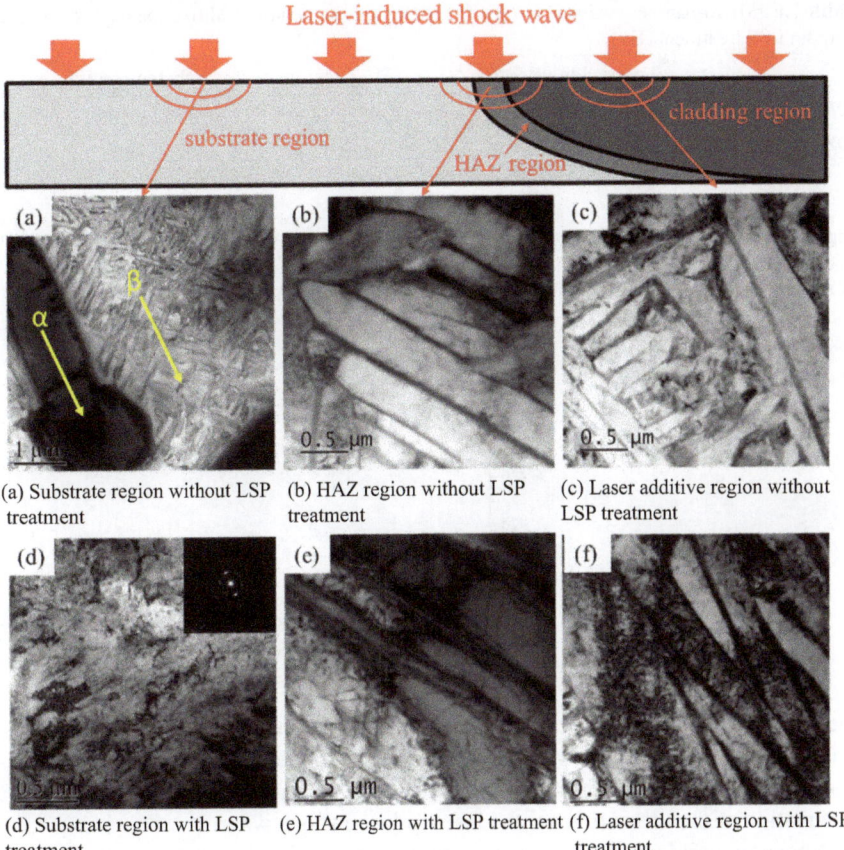

Fig. 7.17 Surface TEM photographs in different regions of the laser additive manufactured specimen with and without LSP treatment

7.3.4 Mechanical Properties

The Vickers micro-hardness tests were performed on an MVS–1000JMT2 micro-hardness tester, using an indentation load of 100 g at the transverse section, a dwell time of 10 s. For each region of the specimen, the hardness value was regarded as an average of 10 measurement results and a confidence interval of 95%. The measured Vickers micro-hardness values for the different regions of the laser additive specimens with and without LSP treatment are listed in Table 7.6. The micro-hardness increased from 350 ± 9.6 Hv of the original specimen to 415 ± 19.7 Hv of the laser additive region. And both of the surface micro-hardnesses in the three regions increased after LSP treatment. Further, the cross-sectional micro-hardness in the laser additive region was measured and the results are shown in Fig. 7.18. According to the process of the preparation of the specimens (Fig. 7.11), the through-depth in

Table 7.6 Micro-hardness values for the different regions of the laser additive specimens with and without LSP treatment (Hv)

	Laser additive region	HAZ region	Substrate region
Without LSP	415 ± 19.7	390 ± 16.8	350 ± 9.6
With LSP	470 ± 11.5	430 ± 12.4	402 ± 13.7

Fig. 7.18 Cross-sectional micro-hardness in the center of the laser additive specimen after LSP treatment

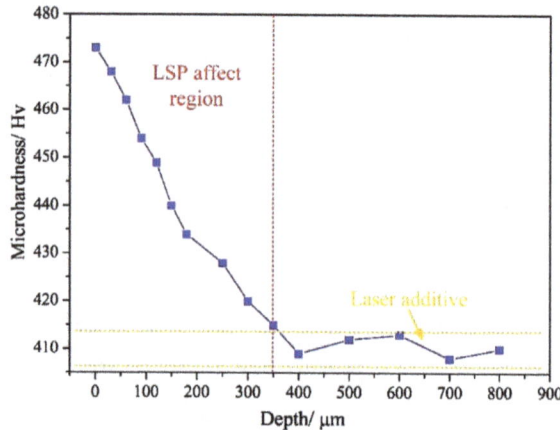

the center of the specimen was of laser additive material. This is why the substrate micro-hardness was about 410 Hv. After LSP treatment, the affected depth was about 350 μm, and the gradient change of micro-hardness was consistent with the attenuation of the pressure from the laser-induced shock wave.

Room temperature tensile properties of the LAMed Ti alloy specimens before and after LSP treatment are shown in Table 7.7. Figure 7.19 shows the typical engineering stress–strain curves for different samples. Both the strength and the ductility of the specimen after LSP treatment are superior to that of the laser additive specimen. The tensile strength with additive increased by 11% to 1058 MPa compared with 953 MPa before LSP treatment, and the yield strength increased from 896 to 962 MPa, by 7.3%. The elongation rate of the laser additive specimen noticeably increased from 4.3 to 6.2% after LSP treatment. The brittle and local tensile stress in the laser additive region led to the pre-failure of the laser additive specimens. These results indicate that LSP can effectively eliminate the adverse effects produced by LAM and improve the tensile strength, yield strength and the elongation rate of LAM specimens.

Table 7.7 Room temperature tensile properties of the laser additive manufactured Ti alloy

Samples	UTS/MPa	YTS/MPa	EL/%
Without LSP	953	896	4.3
With LSP	1058	962	6.2

Fig. 7.19 Room
temperature tensile
engineering strain–stress
curves for different samples

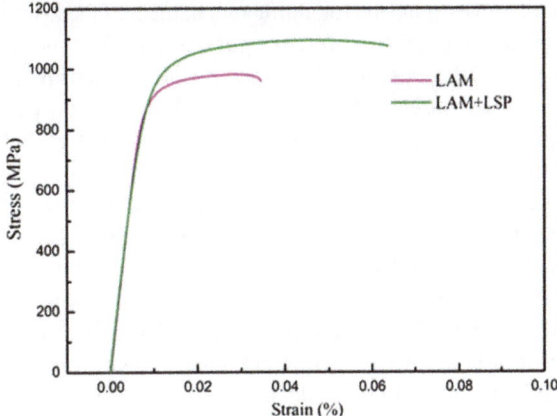

7.3.5 *Fatigue Strength*

The fatigue tests show that all three types of specimens failed at the middle of the specimen, and fatigue cracks initiated at the surface of the laser additive layer. The fatigue strength deduced from the step-loading method at 2×10^5 are shown in Table 7.8, the fatigue strength of the substrate specimens and laser additive specimens were about 401 MPa and 365 MPa, respectively. The fatigue strength of laser additive specimens decreased by 9% compared with the substrate specimens. Namely, LAM could repair the geometrical damage, however it could not regain the fatigue strength. In comparison, the results of the fatigue tests clearly show that the LSP post-laser additive treatment had a strong impact on fatigue behavior. The fatigue strength of LSP post-laser additive specimens improved to 451 MPa, increasing by 12.5% compared with the substrate specimens and 23.6% compared with the laser additive manufactured specimens.

Table 7.8 The fatigue process and results of the three types of specimens

Specimen state	σ_0/MPa	Step	$\Delta\sigma$/MPa	N_f/cycles	σ_{FS}/MPa	Average values
Substrate	400	2	40	41,396	408	401
	300	5	30	39,940	396	
	300	5	30	67,426	400	
LAMed	300	3	30	180,945	357	365
	300	4	30	88,154	373	
	300	4	30	32,765	364	
LAM + LSP	300	8	30	19,864	483	451
	350	4	35	115,240	440	
	350	4	35	55,902	430	

To investigate the regaining mechanism of the laser additive manufactured Ti alloy subjected to LSP, the fracture surface morphologies of all three surface states were observed through SEM. The specimens chosen were fractured at 410 MPa (substrate), 373 MPa (laser additive) and 430 MPa (LSP post-laser additive) (Fig. 7.20). The quasi-cleavage fracture with a typical river pattern and a primary transgranular fracture were observed. The fatigue crack initiation (FCI) of the substrate specimen occurred at the surface (Fig. 7.20a). From low-magnification SEM observation, the crack initiated from the surface region of the laser-additive specimen (Fig. 7.20b), and multiple FCI locations were found (Fig. 7.20g), which resulted in reduced fatigue strength compared with the substrate specimen that failed from a single FCI location. The obvious difference was that the FCI location changed to the substrate for the LSP post-laser additive specimen at a high magnification (Fig. 7.20h). The change may be attributed to the introduction of compressive residual stress produced by LSP. In addition, plenty of dimples were generated on the final fatigue fracture morphologies, and the fracture showed the similar characteristics with a ductile material [18]. The smaller dimples in the final fracture region were found in the laser additive (Fig. 7.20e) and in the LSP post-laser additive (Fig. 7.20f) specimens compared with the substrate specimen (Fig. 7.20d), which indicates that the laser additive specimen had a lower plasticity compared with the substrate. The brittle laser additive layer resulted in reduced fatigue strength. After LSP treatment, the dimples had slightly increased and more tear ridges were found compared with the laser additive specimen.

The reduction in fatigue strength was due to the presence of tensile residual stress (Fig. 7.13a, c)after the LAM process. High tensile residual stresses caused faster

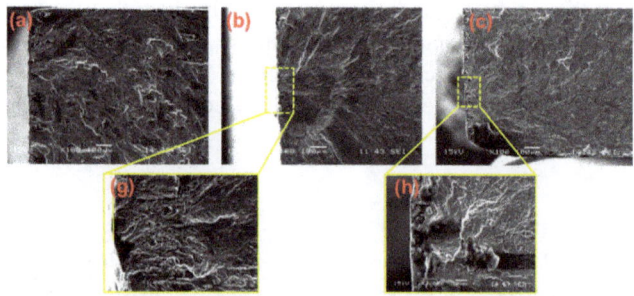

(a) FCI of the substrate with 410 MPa (b) FCI of the laser additive with 373 MPa (c) FCI of LSP post-laser additive (g) The high magnification morphology in FCI of the laser additive (h) The high magnification morphology in FCI of the LSP post-laser additive

(d) The final region of the substrate (e) The final region of the laser additive (f) The final region of the LSP post-laser additive

Fig. 7.20 Fracture surface morphologies

initiation and propagation of cracks through an increase in the effective mean stress during fatigue cycling, resulting in lower fatigue strength [19]. As a post treatment, the laser-induced shock wave produced a gradient in plastic deformation in depth from the top surface, which resulted in the generation of a gradient residual stress. It was noted that there were also high-density dislocations in the grains of the laser additive region and refined grain in the surface layer of substrate region after LSP treatment (Fig. 7.17). Theoretically, the strength of metallic alloys could be predicted by some factors, e.g. the reduction in the size of the grain and an increase in the density of dislocation with plastic deformation [20].

$$\sigma = \sigma_0 + kd^{-1/2} + \alpha Gb\rho^{1/2} \qquad (7.2)$$

where σ is the strength, σ_0 is a friction stress, k is the Hall–Petch constant, d is the grain size, α is a constant, G is the shear modulus, b is the Burgers vector and ρ is the density of dislocation. After LSP treatment, high dense dislocations and some sub-structure at the surface led to the high work hardening and strength. The high surface work hardening created a barrier and restricted the movement of dislocations to the surface. Thus, the required cycles, which lead to the initiation of cracking, were much higher for the LSP post-laser additive specimens than that for the laser additive specimens. On the other hand, the high compressive residual stresses introduced by LSP in the near-surface layer of laser additive specimens reduced the effective mean stress during fatigue cycling during fatigue loading [12]. Thereby, the compressive residual stress prevented or at least delayed the initiation of cracks and reduced the rate of the propagation of cracks [21]. In a word, the increase in fatigue strength of LSP post-laser additive specimens was due to the combined effects of the high magnitude of compressive residual stresses and microstructural changes, i.e. increased density of dislocation and refined grains.

7.4 Enhance Aluminizing via LSP-Induced Gradient Microstructure with Good Thermal Stability

7.4.1 Experiments and Methods

K417 Ni-based cast alloy, mainly consisting of γ solid solution, strengthening phase γ' and ($\gamma + \gamma'$) eutectic, is used as the experimental material in this section (shown in Fig. 7.21a). The presence of the γ and γ' phases was further justified by the TEM observation, as shown in Fig. 7.21b, c. The nominal chemical composition is listed in Table 7.9 and the atomic scale structure of the γ and γ' phases can be seen in reference [22]. Due to the presence of Cr of 8.5–9.5%, the K417 alloy has good resistance to oxidation and to corrosion under the premise of ensuring high-temperature mechanical properties. Other alloying elements such as Co, Mo, Mn, Si

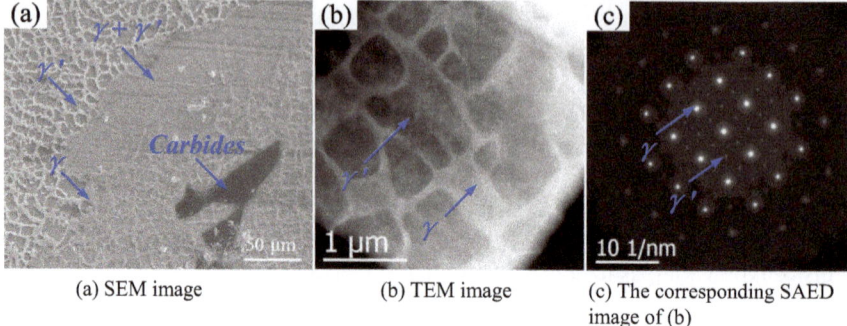

(a) SEM image   (b) TEM image   (c) The corresponding SAED
                                                                  image of (b)

Fig. 7.21 Surface microstructure in the substrate of the K417 Ni-based alloy

Table 7.9 Composition of the K417 cast nickel-base superalloy

Composition	Cr	Co	Mo	Ti	Al	C
Percentage (wt.%)	8.5–9.5	14–16	2.5–3.5	4.5–5.7	4.8–5.7	0.13–0.22
Composition	V	B	Zr	Mn	Si	Ni
Percentage (wt.%)	0.6–0.9	0.012–0.022	0.05–0.09	<0.5	<0.5	Bal

function mainly as solution-strengthening elements and improve the performance of the materials [23].

In the LSP treatment, a laser with the following listed parameters was used: energy is 10.8 J, pulse width is 20 ns, the laser beam diameter is 3.4 mm, and the density of power is 6 GW/cm^2. Two-sided LSP was used on the standard vibration specimens. The paths of the laser spots and the LSP region are demonstrated in Fig. 7.6.

Both the samples pre-treated or pre-untreated by LSP treatment were washed carefully by acetone and pure alcohol and then were immediately subjected to a gas-aluminizing treatment in a homemade device, as shown in Fig. 7.22. The experimental materials were embedded in a birdcage clamp and the gas-aluminizing process was carried out in a special pit furnace with a pre-evacuated vacuum mechanical pump. Al–Fe powder, ammonium chloride (NH$_4$Cl), and aluminum oxide (Al$_2$O$_3$) powder were used as infiltration agent, activator and filler, respectively. Subsequently, the samples were aluminized under a high-temperature condition. The furnace was vacuumed to 10^{-3} mbar and then filled with argon to avoid surface oxidation and ensure inert atmosphere conditions during the length of the whole process. The aluminizing temperature was 900°C + 10 °C and the holding time was 2 h.

MFS-7000 XRD equipment with Cu-kα radiation ($\lambda = 1.5406$ nm) was adopted to characterize the phase analysis of aluminide coatings with a voltage of 50 kV, a current of 40 mA and a take-off angle of 6°. The diffraction data were collected over a 2θ range of 20°–100°and step width was 0.02°. MDI Jade 6 software was used to determine the sizes of the grain and the integrated areas for β-phase as well as the δ-phase diffraction peaks in the XRD spectra.

Fig. 7.22 Schematic representation of the gas-aluminizing device

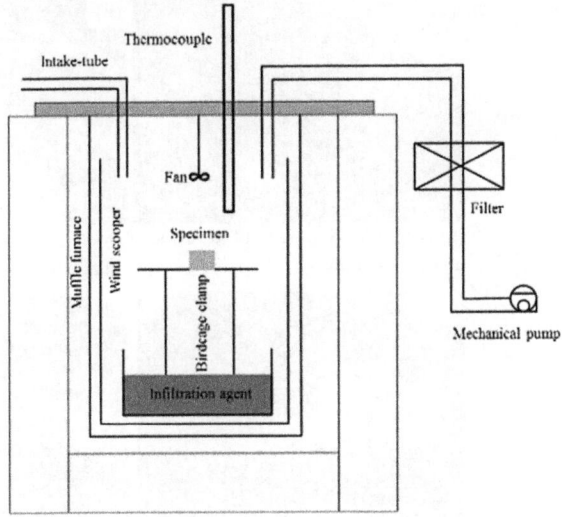

To precisely characterize the surface microstructure of the K417 Ni-based alloy after LSP treatment, thin plates with a thickness of about 20 μm were chosen and TEM observation was adopted. TEM was performed using JEM 2100F with the experimental parameters: FEG (field emission gun): 200 kV; point resolution: 0.23 nm and line resolution: 0.14 nm. A JEOL/JSM-6360LV SEM with a voltage of 20 kV was employed to observe the microstructural morphological images and the compositions within the aluminizing coatings was determined by an equipped energy dispersive spectroscopy (EDS) system.

The D-300-3 electric vibration system was used to design and carry out vibration fatigue tests at first order resonant frequency, room temperature and in air. Figure 7.6 shows the size and shape of the fatigue specimens. The test method is similar to that in Sect. 7.2.4.

7.4.2 Microstructure Evolution Induced by LSP

The typical TEM images at different depths of the K417 Ni-based alloy sample after LSP treatment are shown in Fig. 7.23. Figure 7.23a, b demonstrate surface nanocrystallines produced by LSP treatment with sizes of 30–150 nm. The nanocrystallines with random orientations are indicated by the corresponding circles of selected area electron diffraction (SAED) patterns. From the TEM images, the nanocrystallines are noticeably elongated in the direction perpendicular to that of the LSP process. It is known that during the shock wave propagation, the pressure on the material decreases with an increase in the depth. At a depth of ~2 μm from the treated surface, dislocation cells and high-density dislocations were generated (Fig. 7.23c). Lower but still significant density dislocations are found as the depth further increases to 5 μm,

(a) Bright-field image of the surface   (b) The corresponding dark-field
nanocrystalline   image of image (a),

(c) Dislocation cells and high dense   (d) Dislocations at a depth of 5 μm
dislocations

Fig. 7.23 Typical microstructures at different depths of the K417 Ni-based alloy after LSP treatment

(Fig. 7.23d). It is noticed that the dislocation activity caused by the laser-induced shock wave is the main reason for the formation of surface nanocrystals. The comprehensive explanation of the mechanism of surface nanocrystallization induced by LSP was described in detail in a previous study [5].

The nanocrystalline and the high-density dislocations in the surface layer of the Ni-based alloy produced by LSP can introduce a large number of defective grain boundaries and the aluminizing process was carried out under 900 °C high-temperature conditions. Thus, the thermal stability of the surface gradient microstructure induced by LSP is the key issue. One piece of research showed that at 900 °C, the surface nanostructure can maintain a good thermal stability for a long time [6]. In particular, the largest grain size in the surface nanostructure layer increased from 140 to 220 nm at the annealing temperature of 900 °C, and there still exists some of the smaller grains with a size of around 50 nm. The reasons for the good thermal stability of the surface nanostructure induced by LSP are the lower cold work hardening rate of LSP, the large volume fraction of grain boundaries present in the surface nanocrystalline layer and the stability of the high-density dislocations. The high thermal stability, along with the density of dislocation, provides suitable conditions and plays an active part in high-temperature gas aluminizing.

7.4.3 Aluminizing and Microstructure Characterization

To investigate the effect of the surface nanostructure induced by LSP on the formation of the aluminide coatings, the XRD peaks of the surface coatings obtained with different treatments were analyzed and the results are shown in Fig. 7.24. The surface coatings in the aluminizing and LSP + aluminizing samples both contain strong diffraction peaks relative to the β-NiAl phase, as well as weak diffraction peaks associated to the δ-Ni_2Al_3 phase. According to the "high activity" formation mechanisms [24], it may be concluded that the diffusion layers are primarily developed by inward diffusion of Al inside the Ni matrix, rather than outward diffusion of nickel, which leads to the formation of the dense β-NiAl phase. Ehtemam Haghighi et al. [25, 26] proposed a method for calculating the volume factions of alloy phases based on the integrated areas of diffraction peaks. Due to the existence of two phases in several diffraction peaks, it is difficult to estimate the quantitative volume factions of β-NiAl phase and δ-Ni_2Al_3 based on this method. But we concluded that there was more Al element diffusion into the matrix for the sample treated by pre-LSP based on the integrated areas of diffraction peaks obtained by MDI Jade 6 software.

In addition, the (FWHM and the corresponding location of different samples are shown in Table 7.10. The results show that the FWHM of aluminizing samples with pre-LSP treatment is broader compared with that of aluminizing samples without pre-LSP treatment. Based on the XRD patterns, the grain size of LSP + aluminizing sample changed from several micrometers of the aluminizing-only sample to 32 nm. Meanwhile, the locations of the diffraction peaks showed only small deviations between the two samples. The small deviations of the locations of the diffraction peaks were mainly due to the thermal stability of the micro-strain and nanoscale grains introduced by pre-LSP treatment.

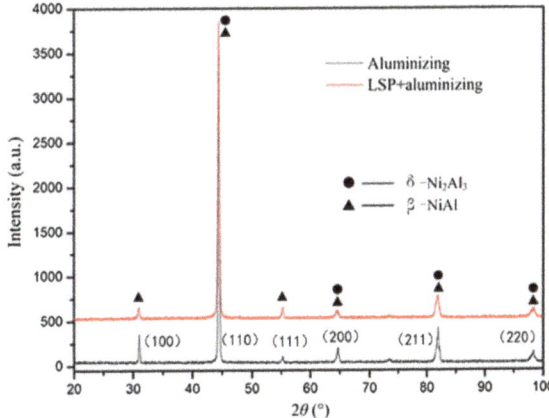

Fig. 7.24 The XRD patterns of the K417 Ni-based alloys after aluminizing processing with and without pre-LSP treatment

Table 7.10 The FWHM and the corresponding locations of the aluminizing and LSP + aluminizing samples

hkl	FWHM/2θ	
	Aluminizing	LSP + aluminizing
100	0.198/31.04	0.314/30.98
110	0.236/44.50	0.279/44.46
111	0.251/55.26	0.292/55.22
200	0.325/64.72	0.458/64.66
211	0.394/81.92	0.547/81.90
220	0.455/98.32	0.525/98.28

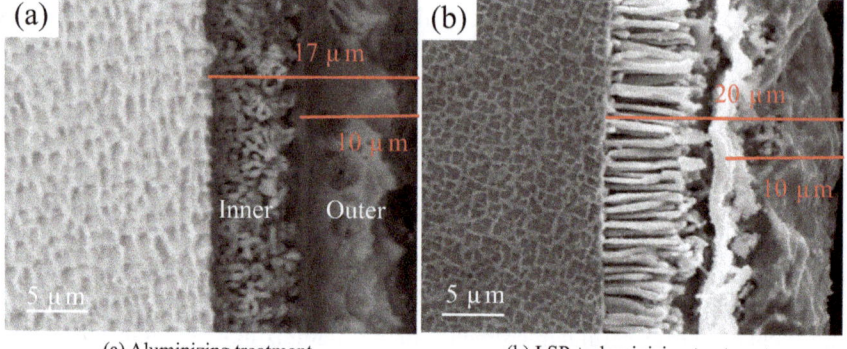

(a) Aluminizing treatment   (b) LSP + aluminizing treatment

Fig. 7.25 SEM images of the structure of the K417 Ni-based alloy aluminizing diffusion layer under different surface treatments

To further analyze the difference between the aluminizing and LSP + aluminizing samples, the SEM observation was adopted to obtain the cross-sectional microstructures (Fig. 7.25). Three layers, referred to as inner, middle and outer, could be found in the aluminide coatings of both samples. Figure 7.25a shows the cross-sectional microstructure of the aluminide coating in the original aluminizing sample. The thickness of the diffusion coating was ~17 μm. The thickness and microstructural characteristics of the diffusion coating in LSP + aluminizing sample were different. During the high temperature aluminizing processing, the pre-produced Al_2O_3 on the material surface can hinder the diffusion of Al atoms into the material. On the other hand, due to the stable β-NiAl phase on the surface, there was no change in the thickness of the outer layer between the aluminizing sample and the LSP + aluminizing sample (Fig. 7.25b), i.e. the thickness of the outer layer of the two samples was always ~10 μm. The random distribution microstructure in the diffusion layer of the aluminizing sample has a tendency to transform into an orderly distribution, with a preferential growth direction and the formation of an acicular structure, in the LSP + aluminizing sample (Fig. 7.25b). This is because the severe plastic deformation induced by LSP provides the energy required for the acicular crystal growth. At the beginning of the aluminizing processing, the locations of the grain nucleation of the β-NiAl and δ-Ni$_2$Al$_3$ phases were formed on the surface of the nickel and chromium atoms, respectively. Under the same aluminizing conditions, the refined nickel particles formed during the LSP treatment significantly increased the rate of NiAl nucleation and shortened the distance between different NiAl nucleation sites. In other words, the formation of surface nanostructure and dislocation defects could effectively shorten the formation time of the dense NiAl layer. In addition, the refined nickel particles also increased the rate of diffusion of Al into the matrix, resulting in the rapid formation of the NiAl layer, which results in an increase in the thickness of the inner layer from ~7 to ~10 μm.

7.4.4 Element Distribution in the Aluminizing Coatings

The cross-sectional SEM images and related EDS results at different depths of different samples are shown in Fig. 7.26. At the beginning of the aluminizing process, the Al diffusion coefficient was orders of magnitude larger than that of Ni [27]. This means that the vapor phase deposited Al diffused into the matrix and reacted with Ni, resulting in the formation of the δ-Ni$_2$Al$_3$ phase. With the increase in temperature and stabilizing heating treatment, the metastable δ-phase (Ni$_2$Al$_3$) further transformed into the stable dense β-phase (NiAl), these could be demonstrated by the increase in Al content at point 1 (Fig. 7.26a) of the outer layer (Fig. 7.26b) and XRD results (Fig. 7.26). When the temperature increased to 900 °C, the diffusion coefficient

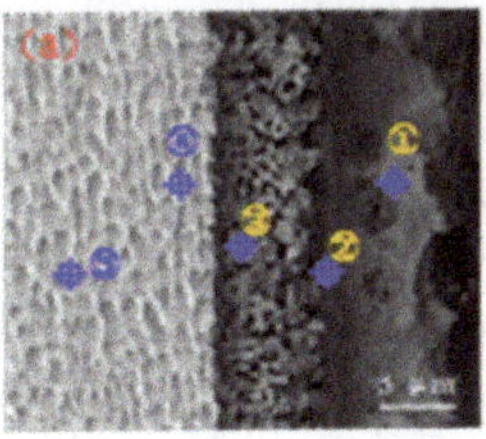

(b) wt. %

Location	Cr	Co	Mo	Ti	Al	C	O	Ni
5	6.56	11.37	2.43	4.28	3.78	11.36	3.22	57.00
4	11.05	15.30	4.03	2.43	3.15	11.11		51.78
3	31.33	16.96	13.77	3.52	5.39		5.85	22.84
2	13.42	5.48	3.44	19.15		44.06		11.93
1	1.56	12.06			35.26		3.39	47.74

(a, b) Aluminizing sample

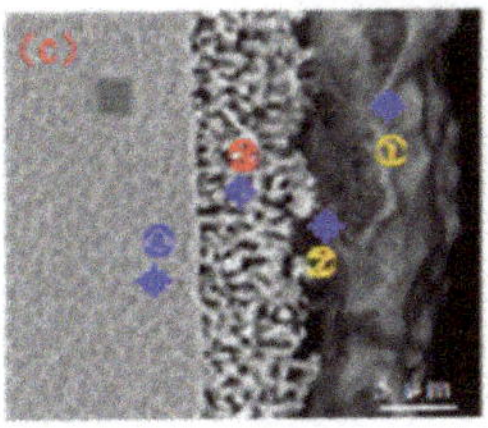

(d) wt. %

Location	Cr	Co	V	Ti	Al	C	O	Ni
4	7.98	12.01	1.45	4.64	5.13	4.58		61.10
3	34.35	17.09	1.29	1.77	2.05	4.41	4.97	18.59
2	5.86	11.1		4.15	26.86	6.41		18.59
1		6.41			31.19	8.33	2.70	51.38

(c, d) LSP + aluminizing sample

Fig. 7.26 SEM cross-sectional images and associated element distributions at different depths of the aluminizing layer of the K417 Ni-based alloy

of Ni also increased significantly, but the δ-Ni$_2$Al$_3$ layer, especially the dense β-phase layer, formed previously, prevented the outward-diffusion of Ni. Therefore, the outward-moving Ni reacted with Al, which diffused across the δ-Ni$_2$Al$_3$/β-NiAl layer, resulting in the formation of the γ'-Ni$_3$Al phase, as seen from the distribution of the element at point 3 (Fig. 7.26a) of the inner layer (Fig. 7.26b). It is worth noting that a "low activity" process also exists in the aluminizing process. According to the "low activity" mechanism [27], the alloy elements, such as Ti, Cr and Co, will precipitate at the interface between the β-NiAl layer and the substrate, and then react with the outward diffusing C element, forming an layer of interdiffusion containing some carbides, such as M$_{23}$C$_6$ and M$_7$C$_3$. This was proved by the EDS result at point 2 of the middle layer in Fig. 7.26b. The presence of a large amount of alloy elements could further restrict the outward diffusion of Ni and promote the formation of the γ'-Ni$_3$Al phase, which contributes to increasing the fatigue property during the cycle loading. Due to the outward-diffusion of alloy elements, the contents of Cr, Co and Mo decreased at the substrate (point 5 in Fig. 7.26b) and increased at and near the inner layer (point 3 and 4 in Fig. 7.26b) compared with that in original samples (Table 7.9).

7.4.5 Mechanism of Diffusion on a Surface of a Nanostructured Layer

Nanostructured materials have numerous grain boundaries and an enhanced diffusivity compared to the original materials because of the lower activation energy at grain boundaries [28]. Moreover, in a previous work [6], we found that the surface nanostructure induced by LSP in K417 Ni-based alloy can keep a good thermal stability even at a high temperature. Therefore, the sample with pre-LSP treatment can provide a large number of fast diffusion channels during the high-temperature gas aluminizing process. In addition, the number of nucleation sites is evidently higher in nanostructured materials, due to the large amount of grain boundaries and high-density dislocations. These are the main reasons for which the β-NiAl phase and a small amount of the δ-Ni$_2$Al$_3$ phase on the surface were found at point 1 of the outer layer (Fig. 7.26c, d): a larger amount of Al atoms diffuse into the substrate compared to the original sample. Underneath the diffusion layer (point 4 in Fig. 7.26c), the contents of Ni and Al were higher than they were at the same location of the original aluminizing sample. This indicates that some Al atoms are capable of diffusing across the surface of the NiAl layer due to LSP-induced surface nanocrystallization. Moreover, carbides also existed in the interdiffusion layer, and the contents of Cr and Co were higher in the inner layer, as seen in point 3 (Fig. 7.26b, d). The increase of Al and Cr contents had a positive role on resistance to oxidation [29]. On the other hand, the presence of a large amount of alloy elements restricted the outward diffusion of Ni, which is why there was a higher Ni content at point 4 (Fig. 7.26d) compared with that at the same point of the original aluminizing sample (Fig. 7.26b).

Accordingly, the SEM micrographs and related EDS results indicated that there existed three layers in both the original aluminizing and LSP + aluminizing samples. The outer layer was formed by the δ-Ni_2Al_3 and Al-rich β-NiAl phases; the middle layer consisted of β-NiAl, enriched by some alloy elements; the inner layer consisted of γ'-Ni_3Al. The difference between the original aluminizing and LSP + aluminizing samples was only in the distributions of the element in the three layers. Therefore, the mechanism of diffusion during the two aluminizing processes could be described in the same manner.

According to the results of Goward [30], the rates of the diffusion of Al atoms in the δ-Ni_2Al_3 phase are three to four orders of magnitude higher than that in the γ'-Ni_3Al and β-NiAl phase at the low temperature, so the inward diffusion of the active Al atoms inside the sample surface occurs, leading to the formation of metastable δ-Ni_2Al_3. The metastable δ-phase (Ni_2Al_3) further transformed into the stable dense β-NiAl phase during the heating treatment. This is why the outer layer consists of δ-Ni_2Al_3 and Al-rich β-NiAl phases. For a short time, further inward aluminum diffusion must occur, and the aluminum content of the advancing interface must decrease to a value at which formation of the β-NiAl by the Al diffusion can no longer occur at some point (middle layer). At the high-temperature condition, the coefficient of Ni diffusion increased significantly. The large amount of outward-moving Ni reacts with the inward diffusing Al atoms, and then promotes the formation of γ'-Ni_3Al in the inner layer (point 3 in Fig. 7.26).

7.4.6 HCF Strength

To further verify the effects of different surface treatments on the fatigue strength of the K417 Ni-based alloy, the high-cycle fatigue test was designed and adopted according to the Chinese technology standard (HB5277-84). Thirty standard vibration fatigue specimens were machined into the dimensions shown in Fig. 7.6. Half of the prepared specimens were treated by aluminizing and the other half was treated by LSP + aluminizing.

The fatigue strength of the K417 Ni-based alloys with different surface treatments is shown in Fig. 7.27. Research on the effect of LSP treatment on the fatigue strength of K417 Ni-based alloys [7] showed that the fatigue strength was increased from 110 MPa, relative to the original samples, to 285 MPa after LSP treatment. The improvement in fatigue strength was attributed to the high amplitude and larger depth compressive residual stress and to microstructural change. In addition, the fatigue strength of the LSP-treated K417 Ni-based alloy was still increased to 230 MPa after a 10 h heating treatment at 900 °C. Although most of the compressive residual stress generated by LSP was relaxed at the high temperature, the additional increase in fatigue strength was mainly attributed to the surface nanostructure induced by LSP with relatively good thermal stability at this temperature.

In this work, the fatigue strength of the sample not subjected to LSP increases from 110 to 226 MPa after aluminizing. The fatigue strength increases to 335 MPa

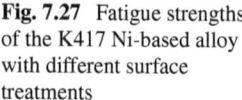
Fig. 7.27 Fatigue strengths of the K417 Ni-based alloy with different surface treatments

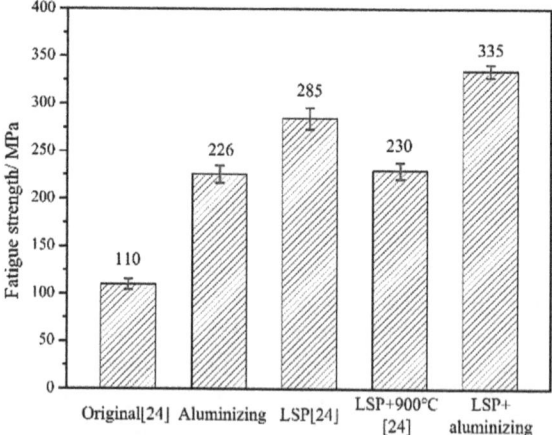

for the aluminizing sample subjected to pre-LSP treatment, which is higher than that of the aluminizing samples (226 MPa) and LSP + 900 °C samples (230 MPa). The mechanism of the improvement of aluminizing with or without pre-LSP treatment on fatigue strength is different from the mechanism of LSP, the detailed discussions are as follows:

It is well known that the γ' phase in Ni-based alloys is the main strengthening phase. During the aluminizing process, some Al atoms could diffuse into the substrate and then react with Ni to form a new γ' phase. Moreover, the diffusion of Ti could replace the Al atoms of the γ'-Ni_3Al phase and form Ni_3(Al, Ti) compounds, which could increase the content of the γ' phase in the surface layer. On the other hand, the addition of the Al element can change the solubility of other alloy elements, such as Ti, Co and Cr. This is why there was a higher content of alloy elements in the diffusion layer (Fig. 7.26). The alloy elements may cause the effect of the strengthening of the solid solution in the diffusion layer, which effectively improves the mechanical properties of Ni-based alloys. The change in the surface microstructure, especially the increase in the γ' phase and the changes in the distribution of the element (mainly the increase in alloy elements) are the main reasons for the improvement in fatigue strength after the aluminizing treatment.

For the LSP + aluminizing samples, the mechanism for improvement is similar to that for aluminizing, but it has some special features. First of all, the SEM observations show that the microstructure in the diffusion layer changes from a random distribution in the original aluminizing samples (Fig. 7.26b) to an orderly distribution, with an acicular structure distribution, in the LSP + aluminizing sample (Fig. 7.26d). Second, the LSP-induced surface nanostructure and high-density defects can provide more channels of diffusion and locations of nucleation, leading to the formation of higher amounts of β-phase and of the γ' strengthening phase and to a thicker layer of diffusion (Fig. 7.26). Third, Xu et al. [31] showed that the compressive residual stress could change the coefficient of the diffusion of alloy elements. The existence of residual stress resulted in more alloy elements diffusing into the surface layer of aluminide of the LSP + aluminizing sample (Fig. 7.26). The increase in alloy elements can increase the content of the solid solution strengthening phase. Moreover, the larger amount of grain boundaries produced by LSP are beneficial for the diffusion of alloy elements (Zr, Ti, Cr, Mn) into the γ' strengthening phase, which act as pinning effects and improve the bonding strength of diffusion layers. The microstructural changes induced by aluminizing is one main reason for the improvement in fatigue strength.

It is worth noting that the surface nanostructure and high-density dislocations induced by LSP had good thermal stability, which has a positive role for fatigue strength. The effect of grain size and dislocations on fatigue strength can be expressed by the Hall–Petch equation [32]:

$$\sigma = \sigma_0 + kd^{-1/2} + \alpha Gb\rho^{1/2} \tag{7.3}$$

where σ_0 is the original strength of the material, k is the Hall–Petch constant, and d is the grain size, α is a constant, G is the shear modulus, b is the Burgers vector and ρ is the density of dislocation. Thus, the thermal stability of surface nanostructure and high-density dislocations at the surface resulted in the high strength for the LSP + aluminizing samples. The high strength created a barrier and restricted the movement of the dislocations during the fatigue load, which resulted in the much higher required cycles to the initiation of cracking. All in all, the improvement in the fatigue strength of LSP + aluminizing samples is due to the increase in Al content and the decrease in grain size induced by the two treatments of LSP and aluminizing.

7.5 Improve High Cycle Fatigue Performance of Gas Tungsten Arc Welded Ti6Al4V Titanium Alloy by Warm Laser Shock Peening

7.5.1 Experiments and Methods

The materials used in this study was a commercially Ti6Al4V titanium alloy plate with a thickness of 3 mm. The heat treatment is annealing at 750–880 °C for 0.5–1 h following by air cooling. The chemical composition of Ti6Al4V titanium alloy is given in Table 7.11. And the basic mechanical properties are shown in Table 7.12. Before welding process, the weld coupons were cleaned using acetone to remove surface dirty and impurities. The titanium alloy plates were welded by WSME-315

Table 7.11 Chemical composition of Ti6Al4V titanium alloy

Component	Ti	Al	V	Fe	C	O	N
Percentage (wt.%)	Bal	6.16	3.95	0.03	0.04	0.06	0.014

Table 7.12 Mechanical properties of Ti6Al4V titanium alloy

Properties	Value
Yield strength $\sigma_{0.2}$ (MPa)	870
Ultimate tensile strength σ_b (MPa)	925
Elongation rate δ (%)	12
Elastic modulus (GPa)	109

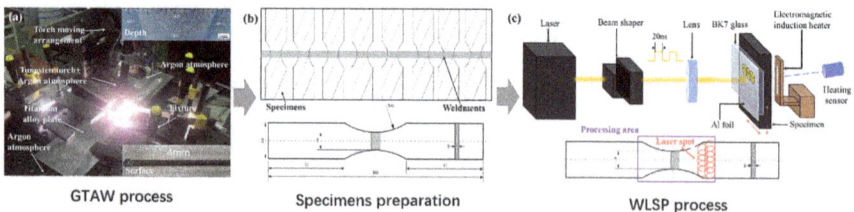

GTAW process   Specimens preparation   WLSP process

Fig. 7.28 Schematic illustration of GTAW process, specimen preparation process and post-welding treatment LSP process

GTAW machine (Riland, China) as shown in Fig. 7.28a. During butt welding, the face and root sides of welding plate were shielded by argon protective atmosphere. Based on our previous trails and error method, the welding parameters were determined as shown in Table 7.13 to make the weldments penetrate the whole plate depth, aiming at simulating the actual weldment with filler wire. From macroscopic morphology of sectioned weldments, it could be clearly seen that weldment was free of internal defects and uniform from the surface to the bottom. According to the microstructure features, it was found the whole weldments was mainly composed of three parts, welding zone (WZ, 0–2 mm from the welding center), heat affected zone (HAZ, 2–4 mm from the welding center) and basal zone (BZ, >4 mm from the welding center). The vibration specimens were cut from the welding plate according to the GBT 26076-2010 standard [33] by wire cut method as shown in Fig. 7.28b. Then the specimens were grinded and polished to remove the weld reinforcement, waiting for post-welding treatment WLSP process.

As for WLSP process, a Nd:YAG laser (Tyrida, China) was used to produce laser beam with a pulse duration of 20 ns. Due to the high temperature environment in WLSP process, water generally used as the confinement layer in room-temperature

Table 7.13 Detailed laser parameters used for the GTAW experiments

Shielding gas	Argon
Welding current (A)	100
Welding speed (cm/min)	25
Gap between electrode tip and base plate (mm)	3
Shielding gas flow rate (L/min)	8–14
Weldment width (mm)	4

Table 7.14 Detailed laser parameters used for the RT-LSP and WLSP experiments

Process	Lapping rate (%)	Spot diameter (mm)	Pulse duration (ns)	Energy (J)	Power density (GW/cm^2)	Laser impacts	Temperature (°C)
RT-LSP	50	2.2	20	4	5.26	1	20
WLSP	50	2.2	20	4	5.26	1	300

LSP (RT-LSP) would be failed at elevated temperature. To solve this problem, BK7 glass (~2 mm thick) was selected as the new confinement layer for WLSP experiment as results of its high shock impedance and high melting point. In our previous work, the optimal peening temperature for WLSPed Ti6Al4V titanium alloy was 300 °C [34]. Therefore, in present work 300 °C was determined as the processing temperature of WLSPed weldments of Ti6Al4V titanium alloy. Al foil (~100 μm thick) was used as ablating layer to protect the surface layer. The other detailed parameters for WLSP process are listed in Table 7.14, which is determined in our previous experiments in terms of the surface integrity. The processing area of the standard vibration is the whole R-shaped region and two-sided WLSP with two laser beams shocking simultaneously is performed to avoid bending in one-sided WLSP. To compare the effectiveness of RT-LSP and WLSP technique, RT-LSP experiment was also conducted. The parameter used in RT-LSP was same to that in WLSP except for temperature, as shown in Table 7.13.

The residual stress analysis of welding plate before and after WLSP was conducted using LXRD diffraction equipment (Proto, Canada) via sin2ψ method. The X-ray source was Cu-Kα and the X-ray beam diameter was 2 mm. The diffraction plane and angle were {213} and $2\theta = 142°$, respectively. The testing points were selected from the welding center to the base metal zone with an interval of 2 mm. The residual stress in the depth direction was measured by an electrolytic polishing machine with a polishing solution of 90% methanol and 10% perchloric acid. Each point was measured by five times and the average values were adapted.

The surface microhardness of WZ, HAZ and BZ was measured via HX-1000TM/LCD Vicker's indentation tester (Taiming, China). The load was 200 g and holding time was 15 s. The average of five tests was taken for each point.

A TR220 potable roughness instruments (Time, China) was used to measure the surface roughness of weldments. The sensor probe driven by internal precision actuator moved uniformly on the target surface, providing displacement information of probe to processor. Based on the collected data, the surface roughness value was obtained.

A Talos F200s TEM (FEI, USA) was conducted to investigate the dislocations and deformation twin activities on the top surface. The TEM samples were prepared by the focused ion beam (FIB) method via a LYRA3 XMU FIB instrument (Tescan, Czech). The resolution of ion guns was 2.5 nm and it operated at 500 V–30 kV with beam of 1 pA-50 nA. The accelerating voltage of field emission electron gun was 30 kV and its working distance was 9 mm.

Fig. 7.29 Electron vibration
fatigue testing device and
state of specimens in test

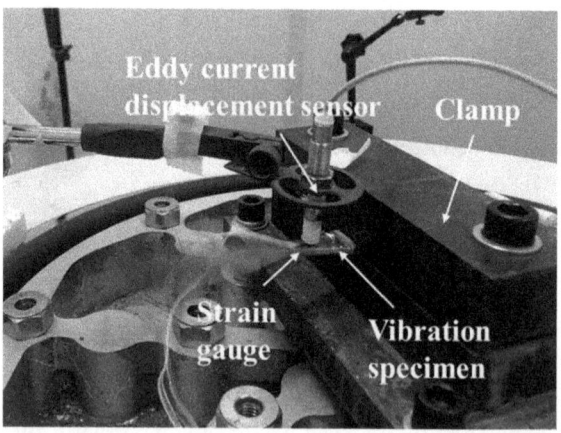

The effectiveness of RT-LSP and WLSP on weldments was confirmed by high cycle vibration fatigue testing using an electron vibration system (Donghua, China). During fatigue testing, a 10^6 cycle number was considered to be the theoretical critical point of fatigue limit. As shown in Fig. 7.29, one end of the vibration specimen was fixed by the clamp, while the other end was free. An eddy current displacement sensor was used to monitor the tip amplitude of the specimens. The vibration stress applied on the center of the notched region was measured by a gauge during the fatigue testing. Before the fatigue test, it was necessary to determine the linear relationship between tip amplitude and the applied vibration stress at low stress level. In this way, the true applied stress could be calculated based on this relationship when the gauge failed at high stress level. When the crack initiated from the specimens, the resonant frequency would change. Thus, once the resonant frequency decreased by 3%, the specimen was defined as failure.

It should be noted that in present study the high cycle fatigue limit of weldments was obtained via the step-loading method [35]. It was a such technique that the fatigue testing started at an initial stress level that the specimens would not fail at 10^6 cycles. After the first 10^6 cycles, the stress was raised into the next loading block by an amount not exceeding 10%. This procedure was repeated until failure occurred. Then the high cycle fatigue limit was given by

$$\sigma_{HCF} = \sigma_{pr} + N_f(\sigma_f - \sigma_{pr})/10^6 \tag{7.4}$$

where σ_{pr} was the maximum stress of the loading prior to the failure, σ_f and N_f were the maximum stress and cycle number of the failure block, respectively. The accuracy of this method has been confirmed in Ti6Al4V titanium alloy [36] and Ni-based superalloy [37, 38] comparable with conventional constant amplitude stress-life method. It could quickly evaluate the high cycle fatigue limit of material. Thus, in this work we would use step-loading technique to evaluate the fatigue limit of weldments with

and without LSP. To assure the reliability of the results, five specimens for each state
were used to test and the average was taken.

7.5.2   Residual Stress Distribution

Figure 7.30 showed the residual stress distribution on the surface of weldments
subjected to RT-LSP and WLSP. It was very clear that the residual stress in the WZ
of the as-welded specimen were compressive, while the area in immediate boundary
of WZ and HAZ developed tensile residual stress. In the HAZ, the tensile residual
stress reached maximum, with an average value of around 175 MPa. In addition, it
was found the width of HAZ from the viewpoint of residual stress almost doubled
that in terms of microstructure features as mentioned in Sect. 7.2.2. Then, with the
increasing distance from the weldment center, the tensile residual stress decreased.
In the BZ far from fusion zone, the compressive residual stress was generated. From
the residual stress profile of weldment, we could know that the tensile residual stress
was mainly existed in and around the HAZ, resulting in detrimental effects to the
fatigue performance of the welded joint. The reason for this phenomenon was unequal
thermal expansion and contraction during heating and cooling cycle in HAZ [39, 40].
After RT-LSP treatment, the original tensile residual stress was converted into high
amplitude compressive residual stress with an average value of around −575 MPa.
Moreover, it is also noted that the compressive residual stress exhibited a relatively
uniform distribution in the whole of weldment, which was beneficial to retard the
crack initiation because crack generally originated from the stress concentration
point. The uniform residual stress field was attributed to the 50% lapping rate in RT-
LSP process. As for the specimens processed by WLSP, it was found that the value
of compressive residual stress on the surface of weldments showed slight decrease as

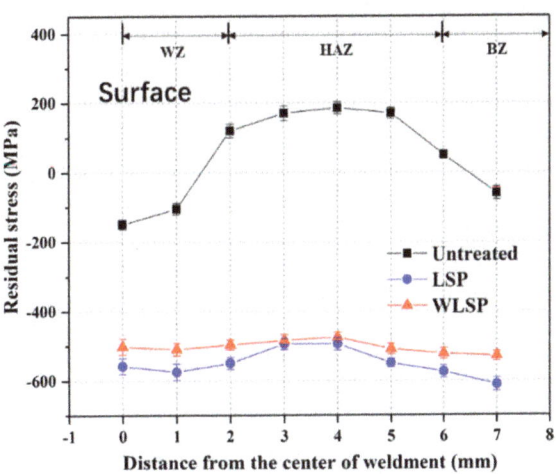

Fig. 7.30 Surface residual
stress distribution from the
weldment center

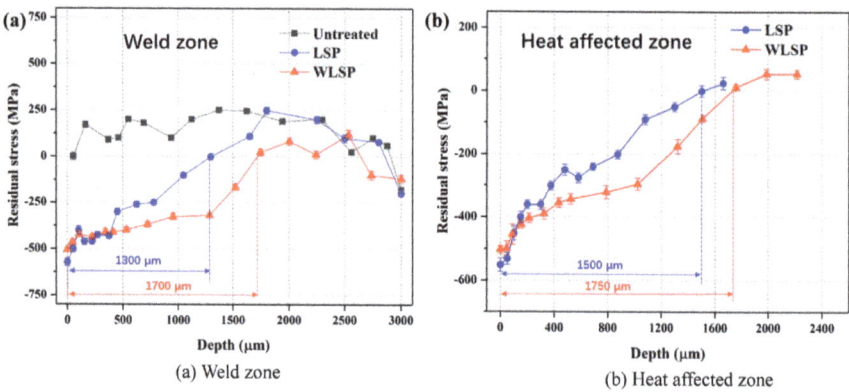

Fig. 7.31 Residual stress distribution in depth

compared with that of RT-LSPed specimens, which was fluctuated around -500 MPa. This was attributed to the thermal effects at the elevated temperatures, making the compressive residual stress release partly.

Figure 7.31 showed the sectioned residual stress profiles in WZ and HAZ. As shown in Fig. 7.31a, the tensile residual stress in the WZ was observed from the welding surface to the bottom in the as-welded specimen. After RT-LSP, high amplitude with a deep affected layer was introduced into the materials. The maximum compressive residual stress located on the surface and then the value of compressive residual stress reduced along the depth direction. This was because the pressure of shock wave attenuated during the propagation process [41–43]. The thickness of compressive residual stress affected layer in WZ reached 1300 μm, while in HAZ it was 1500 μm as shown in Fig. 7.31b. It was found that the cross-sectional residual profiles in HAZ presented the similar distribution like WZ. Therefore, we could conclude that RT-LSP could make the initial tensile residual stress in weldment turn into compressive residual stress with a high-amplitude value and a deeper affected layer. In addition, from Fig. 7.31a, b it was found that WLSP could introduce deeper compressive residual stress layer into the treated materials. For example, in weld zone, the thickness of compressive residual stress layer induced by WLSP has reached 1700 μm, increasing by 30.8% as compared with that of LSPed specimens. The same tendency was also observed in heat affected zone. The thickness of compressive residual stress layer in LSPed samples was 1500 μm, while that in WLSPed specimens increased to 1750 μm (improved by 16.7%). Thus, it could be concluded that high temperature contributed to deeper compressive residual stress layer. The reason for this phenomenon was that the flow stress and Hugoniot elastic limit of materials would decrease under the high temperature environment, which results in more severe plastic deformation under the effects of the same shock wave [44, 45].

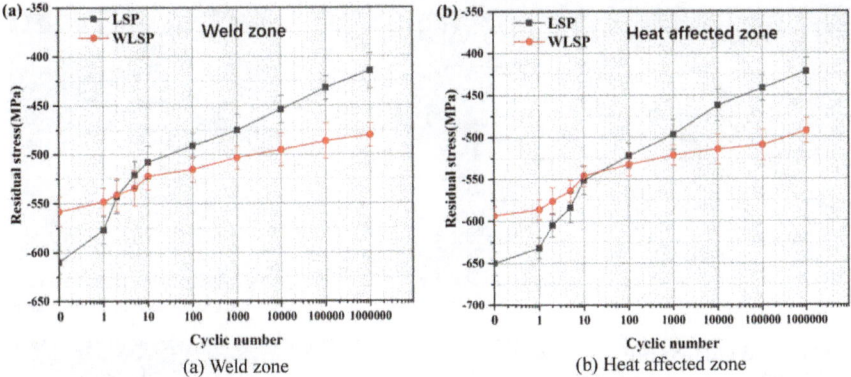

Fig. 7.32   Residual stress relaxation in RT-LSPed and WLSPed specimens after different cycles

7.5.3   Cyclic Stability of Surface Residual Stress

Since the components and parts of aero-engine generally suffered from cyclic vibration stress, the cyclic stability of compressive residual stress induced by WLSP was as important as its magnitude. Figure 7.32 showed the residual stress relaxation in RT-LSPed and WLSPed specimens after cyclic loading at 500 MPa maximal stress. From Fig. 7.32a, it could be observed that although the initial compressive residuals stress of weld zone induced by WLSP was lower than that induced by RT-LSP at room temperature, the releasing rate of compressive residuals stress in WLSPed samples was significantly decreased. For example, after 1000 K cyclic numbers the compressive residual stress from LSP decreased from −610 to −415 MPa, which corresponds to a 32% decrease, while WLSP decreased by only 14% (from −558 to − 480 MPa). Besides, in heat affected zone, the compressive residual stress induced by WLSP also exhibited higher stability. after 1000 K cyclic loading the residual stress magnitude of the LSP samples decreased from −650 to −422 MPa (decrease by 35%), while that of the WLSP samples reduced from −593 to −492 MPa (decreased by 17%), as shown in Fig. 7.32b. Thus, it could be found that the cyclic stability of compressive residual stress induced by WLSP was higher than that generated by RT-LSP, which provides more benefits for the improvement of fatigue performance by delaying crack initiation and decreasing propagation speed [46, 47].

7.5.4   Microstructures Induced by WLSP

Figure 7.33 showed the TEM image of the grain structure in the surface layer of weldments with and without WLSP. As shown in Fig. 7.33a, the original grains of weldments exhibited lath shape without few dislocations, while after WLSP process we could see that the lath-shaped structures broke into several parts under the effects of

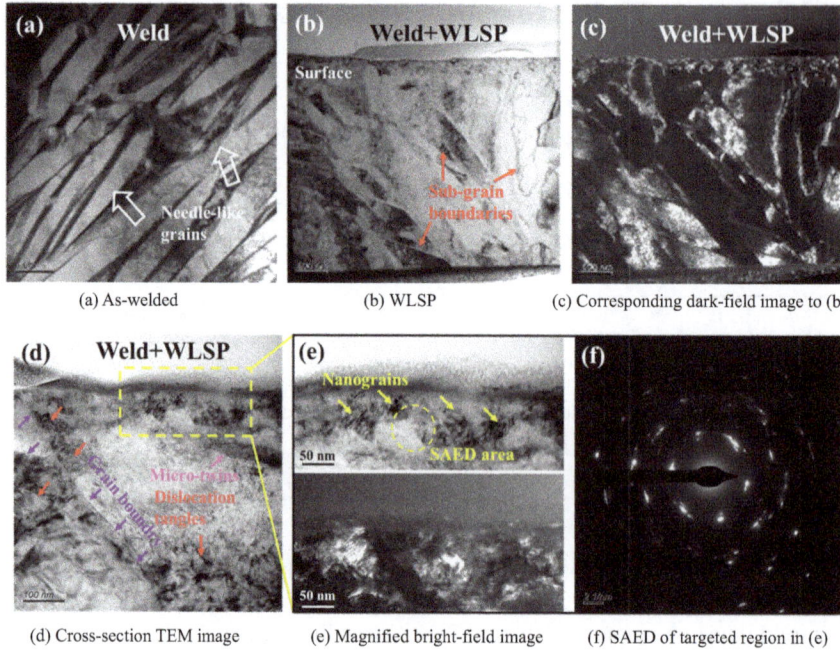

(a) As-welded   (b) WLSP   (c) Corresponding dark-field image to (b)

(d) Cross-section TEM image   (e) Magnified bright-field image   (f) SAED of targeted region in (e)

Fig. 7.33 Grain morphologies of the weldment with and without WLSP

shock wave (Fig. 7.33b). At the same time, high density dislocations were introduced into the subsurface of the materials, as evident in TEM dark image (Fig. 7.33c). From Fig. 7.33d, it could be clearly seen that the dislocation distributed ununiformly and significant dislocation pile-up were performed along the grain boundaries. Due to the resistivity of grain boundaries, massive dislocations accumulated here, contributing to the formation of dislocation tangles. In addition, one micro-twin with a width of 50 nm was observed at the subsurface. On the top surface there were lots of nanograins with the grain size below 100 nm, as was evident in Fig. 7.33e. The selected area electron diffraction (SAED) taken from the area highlighted by yellow line in Fig. 7.33e exhibited a circular shape with discontinuous and elongated diffraction spots as shown in Fig. 7.33f, suggesting the original coarse grains were refined to nanograins with random crystallographic orientations after WLSP process. And the depth of nanograins layer was around 100 nm based on the TEM image.

The microstructure at the subsurface below nanograins layer was shown in Fig. 7.34. It was found in this layer some complex dislocation structures such as dislocation tangles were obviously identified, as shown in Fig. 7.34a, b. From the literatures [48–52] it has been noticed that such dislocation tangles could transform into sub-grains boundaries to separate the original coarse grain, subsequently developed as high angle grain boundaries with the increasing strain and stress. In addition, two micro-twins with a width of around 50 nm was observed as evident in Fig. 7.34c, implying the activity of deformation twins under WLSP. It was generally known that

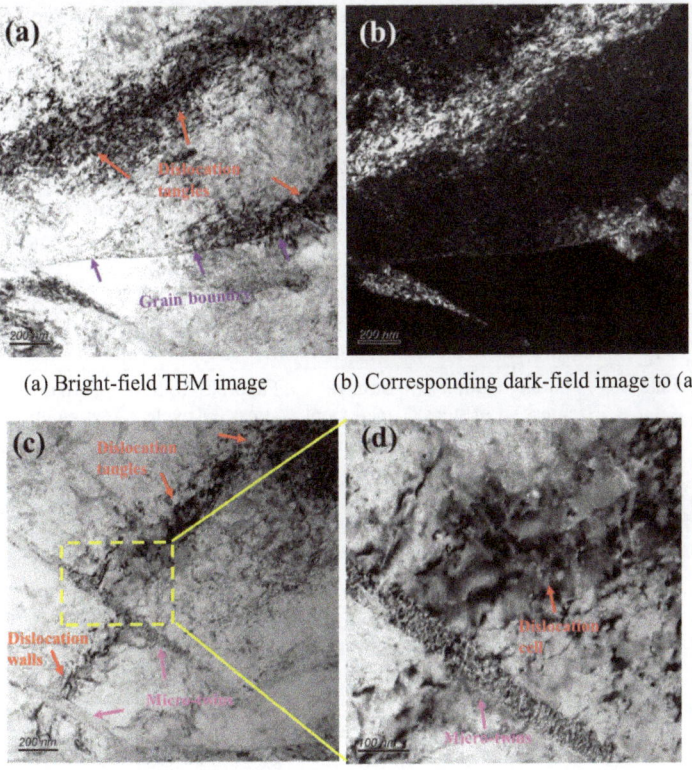

(a) Bright-field TEM image   (b) Corresponding dark-field image to (a)

(c) Intersection of micro-twins and dislocations   (d) Magnified TEM image

Fig. 7.34 Typical deformed structure features at the subsurface below nanograins layer induced by WLSP

when metal materials are subjected to severe plastic deformation at high temperature, the deformation twinning would activate to accommodate the deformation if the dislocation sliding was unable to match the applied strain rate [53]. Also, typical deformed structures such as dislocation wall and dislocation tangles were observed in Fig. 7.34c. The magnified image of the border between micro-twins and dislocations was shown in Fig. 7.34d. We could see that the dislocations couldn't penetrate the twin boundaries and generate dislocation cell structure near the twin boundaries, which helped to improve the local dislocation density.

7.5.5   High Cycle Fatigue Performance

As we know, the sustainability and reliability of the components mainly depended on fatigue limit of materials. To confirm the effectiveness of WLSP process on the fatigue strength of as-welded specimens, high cycle vibration fatigue testing was performed by step-loading method. The initial stress was determined as 350 MPa and corresponding stress increment was 35 MPa. According to Eq. (7.4), the average fatigue limit of five measurements for as-welded specimen was calculated as 399 $\pm$ 5 MPa, as shown in Table 7.15 and Fig. 7.35a. As for the specimens processed by RT-LSP, it was reported in our previous work that the fatigue limit of TC6 [41] and TC11 [8] titanium alloy subjected by RT-LSP could be improved by around 20% compared with untreated specimens. Based on these results, in present work we

Table 7.15 Fatigue testing results of as-welded Ti6Al4V titanium alloy before and after RT-LSP/WLSP by step-loading method

Label	State	Initial stress (MPa)	Number of steps	Stress increment (MPa)	Failure stress (MPa)	Total cycle number ($\times 10^6$)	Fatigue strength (MPa)	Average (MPa)
1	Untreated	350	3	35	420	2.4312	400.0	399 $\pm$ 5
2	Untreated	350	3	35	420	2.2265	392.9	
3	Untreated	350	3	35	420	2.3963	398.9	
4	Untreated	350	3	35	420	2.2549	393.9	
5	Untreated	350	3	35	420	2.6484	407.7	
6	RT-LSP	450	3	45	540	2.9375	537.2	540 $\pm$ 18
7	RT-LSP	450	3	45	540	2.4309	514.4	
8	RT-LSP	450	3	45	540	2.7347	528.1	
9	RT-LSP	450	4	45	585	3.3829	557.2	
10	RT-LSP	450	4	45	585	3.5404	564.3	
11	WLSP	450	3	45	540	2.4054	558.2	568 $\pm$ 19
12	WLSP	450	3	45	540	2.2341	550.5	
13	WLSP	450	4	45	585	3.1267	590.7	
14	WLSP	450	3	45	540	2.2137	549.6	
15	WLSP	450	4	45	585	3.1276	590.7	

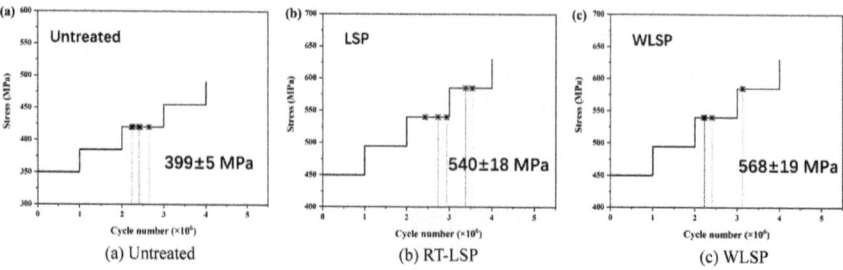

Fig. 7.35 High cycle vibration fatigue limit at 10^6 cycle numbers

determine 450 MPa as the initial stress for as-welded specimen strengthened by RT-LSP. And the fatigue results were shown in Table 7.15 and Fig. 7.35b, we could see that the calculated fatigue limit for RT-LSPed specimen was 540 ± 18 MPa, improved by 35.3%. This result indicated that RT-LSP could effectively improve the high cycle vibration fatigue limit of as-welded specimens. As for WLSPed specimens, it was found its fatigue limit was 568 ± 19 MPa, improved by 42.3% as compared with untreated samples (Fig. 7.35c). This suggested WLSP was more effectiveness on the improvement of fatigue performance of materials.

In fatigue test, large increase in high cycle vibration fatigue limit following WLSP was found. Combining with the results on residual stress, microhardness, surface roughness and microstructure, the strengthening mechanism of weldments subjected to WLSP was discussed in detail. It was analyzed that the following several factors contributed to the improving fatigue limit of weldments:

(I)   It was very clear that the original tensile residual stress in and around HAZ was converted into high amplitude compressive residual stress with an average of around -574 MPa after WLSP and the thickness of compressive residual stress affected layer has reached 1500 μm. As we all know, pre-existing tensile residual stress was extremely harmful to the fatigue performance of materials because it could critically combine with the work tensile residual stress together, causing the enhanced true stress. In contrast, the high near-surface compressive residual stress could counteract part applied stress, making the initiation of cracks transfer to subsurface. However, at the subsurface more restrictions for deformation existed, which helped to prevent or at least delay the initiation of cracks. In other words, the compressive residual stress induced by WLSP could act as a shield from the applied stress while in operation, extending the fatigue life of crack initiation significantly.

(II)   The compressive residual stress induced by WLSP exhibited higher stability with the cyclic loading. This was attributed to the formation of higher dense and more complicate dislocation structures, resulting in an increase in the stress amplitude needed to induce the movement of new mobile dislocations [46]. In this way, the resistance to crack initiation and crack growth was increased by more stable compressive residual stress, thereby leading to higher fatigue performance improvement.

(III) TEM image showed that a nanograin layer with a thickness of 100 nm was generated on the top surface of weldment after WLSP. It has been widely accepted that the nanograins materials could exhibited superior mechanical properties compared with the coarse grain materials [54–56]. For fatigue performance, the nano-scale grains with more grain boundaries could effectively retard the crack initiation by enhanced the strength of materials via suppressing strain localization and dislocation movement, thereby imparting unprecedented resistance to the high cycle fatigue. It was known that the fatigue crack generally initiated from the surface. Thus, the nanograins layer on the top surface could just inhibit the initiation of fatigue crack here. As for the subface, the crack growth could be blocked by the high dense dislocations via restraining the plastic flow [8], making it harder for crack initiation and propagation. In this way, the high cycle vibration fatigue limit of weldments was significantly improved after WLSP.

References

1. S. Luo, L. Zhou, X. Nie, Y. Li, W. He, The compound process of laser shock peening and vibratory finishing and its effect on fatigue strength of Ti-3.5Mo-6.5Al-1.5Zr-0.25Si titanium alloy. J. Alloy. Compd. **783**, 828–835 (2019)
2. R. Fabbro, J. Fournier, P. Ballard, D. Devaux, J. Virmont, Physical study of laser-produced plasma in confined geometry. J. Appl. Phys. **68**(2), 775–784 (1990)
3. H. Suzuki, T. Moriwaki, T. Okino, Y. Ando, Development of ultrasonic vibration assisted polishing machine for micro aspheric die and mold. CIRP Ann-Manuf. Technol. **55**(1), 385–388 (2006)
4. S.H. Luo, L.C. Zhou, X.D. Wang, X. Cao, X.F. Nie, W.F. He, Surface Nanocrystallization and amorphization of dual-phase tc11 titanium alloys under laser induced ultrahigh strain-rate plastic deformation. Materials **11**(4) (2018)
5. S.H. Luo, Y.H. Li, L.C. Zhou, X.F. Nie, G.Y. He, Y.Q. Li, W.F. He, Surface nanocrystallization of metallic alloys with different stacking fault energy induced by laser shock processing. Mater. Des. **104**, 320–326 (2016)
6. S.H. Luo, X.F. Nie, L.C. Zhou, X. You, W.F. He, Y.H. Li, Thermal stability of surface nanostructure produced by laser shock peening in a Ni-based superalloy. Surf. Coat. Technol. **311**, 337–343 (2017)
7. L.C. Zhou, Y.H. Li, W.F. He, G.Y. He, X.F. Nie, D.L. Chen, Z.L. Lai, Z.B. An, Deforming TC6 titanium alloys at ultrahigh strain rates during multiple laser shock peening. Mater. Sci. Eng. A-Struct. Mater. Prop. Microstruct. Process. **578**, 181–186 (2013)
8. X. Nie, W. He, S. Zang, X. Wang, J. Zhao, Effect study and application to improve high cycle fatigue resistance of TC11 titanium alloy by laser shock peening with multiple impacts. Surf. Coat. Technol. **253**, 68–75 (2014)
9. M.A. Meyers, F. Gregori, B.K. Kad, M.S. Schneider, D.H. Kalantar, B.A. Remington, G. Ravichandran, T. Boehly, J.S. Wark, Laser-induced shock compression of monocrystalline copper: characterization and analysis. Acta Mater. **51**(5), 1211–1228 (2003)
10. C. Huang, T.G. Murthy, M.R. Shankar, R. M'Saoubi, S. Chandrasekar, Temperature rise in severe plastic deformation of titanium at small strain-rates. Scripta Mater. **58**(8), 663–666 (2008)
11. X.Q. Zhang, H. Li, S.W. Duan, X.L. Yu, J.Y. Feng, B. Wang, Z.L. Huang, Modeling of residual stress field induced in Ti-6Al-4V alloy plate by two sided laser shock processing. Surf. Coat. Technol. **280**, 163–173 (2015)
12. M.H.E.I. Haddad, K.J. Topper, L.P. Pook, *Metal Fatigue* (Oxford University Press, London, 1974).
13. S.H. Luo, W.F. He, K. Chen, X.F. Nie, L.C. Zhou, Y.M. Li, Regain the fatigue strength of laser additive manufactured Ti alloy via laser shock peening. J. Alloy. Compd. **750**, 626–635 (2018)
14. D.C. Maxwell, T. Nicholas, *A Rapid Method for Generation of a Haigh Diagram for High Cycle Fatigue* (ASTM Special Technical Publication, West Conshohocken, 1999).
15. Y.X. Chew, J.H.L. Pang, G.J. Bi, B. Song, Thermo-mechanical model for simulating laser cladding induced residual stresses with single and multiple clad beads. J. Mater. Process. Technol. **224**, 89–101 (2015)
16. Y.Y. Zhu, J. Li, X.J. Tian, H.M. Wang, D. Liu, Microstructure and mechanical properties of hybrid fabricated Ti-6.5Al-3.5Mo-1.5Zr-0.3Si titanium alloy by laser additive manufacturing. Mater. Sci. Eng. A-Struct. Mater. Prop. Microstruct. Process. **607**, 427–434 (2014)

17. K.Y. Luo, X. Jing, J. Sheng, G.F. Sun, Z. Yan, J.Z. Lu, Characterization and analyses on micro-hardness, residual stress and microstructure in laser cladding coating of 316L stainless steel subjected to massive LSP treatment. J. Alloy. Compd. **673**, 158–169 (2016)

18. J.Z. Zhou, S. Huang, J. Sheng, J.Z. Lu, C.D. Wang, K.M. Chen, H.Y. Ruan, H.S. Chen, Effect of repeated impacts on mechanical properties and fatigue fracture morphologies of 6061–T6 aluminum subject to laser peening. Mater. Sci. Eng. A-Struct. Mater. Prop. Microstruct. Process. **539**, 360–368 (2012)

19. K.F. Walker, J.M. Lourenco, S. Sun, M. Brandt, C.H. Wang, Quantitative fractography and modelling of fatigue crack propagation in high strength AerMet (R) 100 steel repaired with a laser cladding process. Int. J. Fatigue **94**, 288–301 (2017)

20. Z. Zhang, D.L. Chen, Consideration of Orowan strengthening effect in particulate-reinforced metal matrix nanocomposites: a model for predicting their yield strength. Scripta Mater. **54**(7), 1321–1326 (2006)

21. S. Lenders, M. Thone, A. Riemer, T. Niendorf, T. Troster, H.A. Richard, H.J. Maier, On the mechanical behaviour of titanium alloy TiAl6V4 manufactured by selective laser melting: Fatigue resistance and crack growth performance. Int. J. Fatigue **48**, 300–307 (2013)

22. R. Srinivasan, R. Banerjee, J.Y. Hwang, G.B. Viswanathan, J. Tiley, D.M. Dimiduk, H.L. Fraser, Atomic scale structure and chemical composition across order-disorder interfaces. Phys. Rev. Lett. **102**(8) (2009)

23. S.H. Luo, W.F. He, L.C. Zhou, X.F. Nie, Y.H. Li, Aluminizing mechanism on a nickel-based alloy with surface nanostructure produced by laser shock peening and its effect on fatigue strength. Surf. Coat. Technol. **342**, 29–36 (2018)

24. J.W. Lee, Y.C. Kuo, A study on the microstructure and cyclic oxidation behavior of the pack aluminized Hastelloy X at 1100 degrees C. Surf. Coat. Technol. **201**(7), 3867–3871 (2006)

25. S. Ehtemam-Haghighi, Y.J. Liu, G.H. Cao, L.C. Zhang, Influence of Nb on the beta -> alpha martensitic phase transformation and properties of the newly designed Ti-Fe-Nb alloys. Mater. Sci. Eng. C-Mater. Biol. Appl. **60**, 503–510 (2016)

26. S. Ehtemam-Haghighi, Y.J. Liu, G.H. Cao, L.C. Zhang, Phase transition, microstructural evolution and mechanical properties of Ti-Nb-Fe alloys induced by Fe addition. Mater. Des. **97**, 279–286 (2016)

27. F. Bozza, G. Bolelli, C. Giolli, A. Giorgetti, L. Lusvarghi, P. Sassatelli, A. Scrivani, A. Candeli, M. Thoma, Diffusion mechanisms and microstructure development in pack aluminizing of Ni-based alloys. Surf. Coat. Technol. **239**, 147–159 (2014)

28. W.P. Tong, N.R. Tao, Z.B. Wang, J. Lu, K. Lu, Nitriding iron at lower temperatures. Science **299**(5607), 686–688 (2003)

29. J.L. Lv, M. Yang, H. Miura, T.X. Liang, The effect of surface enriched chromium and grain refinement by ball milling on corrosion resistance of 316L stainless steel. Mater. Res. Bull. **91**, 91–97 (2017)

30. G.W. Goward, D.H. Boone, Mechanisms of formation of diffusion aluminide coatings on nickel-base superalloys. Oxid. Met. **3**(5), 475–495 (1971)

31. T.D. Xu, Creating and destroying vacancies in solids and non-equilibrium grain-boundary segregation. Philos. Mag. **83**(7), 889–899 (2003)

32. H.K. Kim, W.J. Kim, Microstructural instability and strength of an AZ31 Mg alloy after severe plastic deformation. Mater. Sci. Eng. A-Struct. Mater. Prop. Microstruct. Process. **385**(1–2), 300–308 (2004)

33. L. Liming, L. Xin, L. Rongfeng, D. Li, N. Xingtao, Metal sheets and strips-axial-force-controlled fatigue testing method, General Administration of Quality Supervision, Inspection and Quarantine of the People's Republic of China and Standardization Administration of the People's Republic of China Beijing (2011)

34. X. Pan, W. He, X. Huang, X. Wang, X. Shi, W. Jia, L. Zhou, Plastic deformation behavior of titanium alloy by warm laser shock peening: Microstructure evolution and mechanical properties. Surf. Coat. Technol. **405**, 126670 (2021)

35. D.C. Maxwell, A rapid method for generation of a Haigh diagram for high cycle fatigue, in *Fatigue and Fracture Mechanics* (ASTM Special Technical Publication, West Conshohocken, 1999), pp. 626–641

36. S. Srinivasan, D.B. Garcia, M.C. Gean, H. Murthy, T.N. Farris, Fretting fatigue of laser shock peened Ti–6Al–4V. Tribol. Int. **42**(9), 1324–1329 (2009)
37. W.J. Ren, T. Nicholas, Effects and mechanisms of low cycle fatigue and plastic deformation on subsequent high cycle fatigue limit in nickel-base superalloy Udimet 720. Mater. Sci. Eng. A-Struct. Mater. Prop. Microstruct. Process. **332**(1–2), 236–248 (2002)
38. W.J. Ren, T. Nicholas, Notch size effects on high cycle fatigue limit stress of Udimet 720. Mater. Sci. Eng. A-Struct. Mater. Prop. Microstruct. Process. **357**(1–2), 141–152 (2003)
39. B. Dhakal, S. Swaroop, Review: Laser shock peening as post welding treatment technique. J. Manuf. Process. **32**, 721–733 (2018)
40. N. Becker, D. Gauthier, E.E. Vidal, Fatigue properties of steel to aluminum transition joints produced by explosion welding. Int. J. Fatigue **139** (2020)
41. X.F. Nie, W.F. He, L.C. Zhou, Q.P. Li, X.D. Wang, Experiment investigation of laser shock peening on TC6 titanium alloy to improve high cycle fatigue performance. Mater. Sci. Eng. A-Struct. Mater. Prop. Microstruct. Process. **594**, 161–167 (2014)
42. T.R. Praveen, H.S. Nayaka, S. Swaroop, Influence of equal channel angular pressing and laser shock peening on fatigue behaviour of AM80 alloy. Surf. Coat. Technol. (2019)
43. A.G. Sanchez, C. You, M. Leering, D. Glaser, D. Furfari, M.E. Fitzpatrick, J. Wharton, P.A.S. Reed, Effects of laser shock peening on the mechanisms of fatigue short crack initiation and propagation of AA7075-T651. Int. J. Fatigue **143** (2021)
44. C.S. Montross, T. Wei, L. Ye, G. Clark, Y.W. Mai, Laser shock processing and its effects on microstructure and properties of metal alloys: a review. Int. J. Fatigue **24**(10), 1021–1036 (2002)
45. J.Z. Zhou, X.K. Meng, S. Huang, J. Sheng, J.Z. Lu, Z.R. Yang, C. Su, Effects of warm laser peening at elevated temperature on the low-cycle Cross Mark fatigue behavior of Ti6Al4V alloy. Mater. Sci. Eng. A-Struct. Mater. Prop. Microstruct. Process. **643**, 86–95 (2015)
46. C.H. Ye, S. Suslov, B.J. Kim, E.A. Stach, G.J. Cheng, Fatigue performance improvement in AISI 4140 steel by dynamic strain aging and dynamic precipitation during warm laser shock peening. Acta Mater. **59**(3), 1014–1025 (2011)
47. C. Liu, D. Liu, X. Zhang, N. Ao, X. Xu, D. Liu, J. Yang, Fretting fatigue characteristics of Ti-6Al-4V alloy with a gradient nanostructured surface layer induced by ultrasonic surface rolling process. Int. J. Fatigue **125**, 249–260 (2019)
48. C.Y. Cui, T.Y. Wan, Y.X. Shu, S. Meng, X.G. Cui, J.Z. Lu, Y.F. Lu, Microstructure evolution and mechanical properties of aging 6061 Al alloy via laser shock processing. J. Alloy. Compd. **803**, 1112–1118 (2019)
49. L. Chen, X.D. Ren, W.F. Zhou, Z.P. Tong, S. Adu-Gyamfi, Y.X. Ye, Y.P. Ren, Evolution of microstructure and grain refinement mechanism of pure nickel induced by laser shock peening. Mater. Sci. Eng. A-Struct. Mater. Prop. Microstruct. Process. **728**, 20–29 (2018)
50. J.Z. Lu, K.Y. Luo, Y.K. Zhang, G.F. Sun, Y.Y. Gu, J.Z. Zhou, X.D. Ren, X.C. Zhang, L.F. Zhang, K.M. Chen, C.Y. Cui, Y.F. Jiang, A.X. Feng, L. Zhang, Grain refinement mechanism of multiple laser shock processing impacts on ANSI 304 stainless steel. Acta Mater. **58**(16), 5354–5362 (2010)
51. J.Z. Lu, K.Y. Luo, Y.K. Zhang, C.Y. Cui, G.F. Sun, J.Z. Zhou, L. Zhang, J. You, K.M. Chen, J.W. Zhong, Grain refinement of LY2 aluminum alloy induced by ultra-high plastic strain during multiple laser shock processing impacts. Acta Mater. **58**(11), 3984–3994 (2010)
52. J. Wu, S. Zou, Y. Zhang, S. Gong, G. Sun, Z. Ni, Z. Cao, Z. Che, A. Feng, Microstructures and mechanical properties of β forging Ti17 alloy under combined laser shock processing and shot peening. Surf. Coat. Technol. **328**, 283–291 (2017)
53. Z.H. Jin, P. Gumbsch, K. Albe, E. Ma, K. Lu, H. Gleiter, H. Hahn, Interactions between non-screw lattice dislocations and coherent twin boundaries in face-centered cubic metals. Acta Mater. **56**(5), 1126–1135 (2008)
54. J. Long, Q. Pan, N. Tao, M. Dao, S. Suresh, L. Lu, Improved fatigue resistance of gradient nanograined Cu. Acta Mater. **166**, 56–66 (2019)

55. J.Z. Long, Q.S. Pan, N.R. Tao, L. Lu, Residual stress induced tension-compression asymmetry of gradient nanograined copper. Mater. Res. Lett. **6**(8), 456–461 (2018)
56. A. Pineau, A.A. Benzerga, T. Pardoen, Failure of metals III: fracture and fatigue of nanostructured metallic materials. Acta Mater. **107**, 508–544 (2016)

9 789811 617492